Edward T. Dixon

# The foundations of geometry

Edward T. Dixon

**The foundations of geometry**

ISBN/EAN: 9783742875211

Manufactured in Europe, USA, Canada, Australia, Japa

Cover: Foto ©berggeist007 / pixelio.de

Manufactured and distributed by brebook publishing software
(www.brebook.com)

Edward T. Dixon

# The foundations of geometry

# THE

# FOUNDATIONS OF GEOMETRY.

BY

EDWARD T. DIXON.

CAMBRIDGE:

DEIGHTON, BELL AND CO.

LONDON: GEORGE BELL AND SONS.

1891

**Cambridge:**

PRINTED BY C. J. CLAY, M.A. AND SONS,

AT THE UNIVERSITY PRESS.

# CONTENTS.

# ERRATA.

p. 12, l. 7, for *CEEC* read *CEFC*.

p. 12, l. 10, for *PE* read *FE*.

p. 15, l. 23, for *ECP* read *ECF*.

p. 18, l. 11, (twice) for 'Helmholz' read 'Helmholtz.'

p. 26, last line but two, erase 'not.'

# PREFACE.

I BELIEVE that the system of geometry I have set forth in this book is logically sound, and that consequently the more it is discussed and criticised, the more firmly will it become established. I shall therefore be very glad to see any criticisms of my views, whether friendly or hostile, either in the public press, or addressed to me privately, at the address given below. But I have already found that, the subject being such a wide one, criticism is apt to become discursive; and with a view to keeping it to the point I would suggest to my critics and opponents in argument that they should consider categorically the following questions :—

(i) Do you accept the requirements I have laid down for a logical definition ? (see p. 21).

(If not, please state which of them you object to, why you object to it, and what you would propose to substitute for it.)

(ii) Do you entertain a mental concept (which I shall call by the name 'direction') such that the assertion " A Vector is a given amount of transference in a given direction, irrespective of the point of departure," is intelligible to you ?

(iii) If so, does not this concept fulfil all the four requirements of my definition of 'direction' ?

(Whether you think these properties are established by Euclid's geometry, or not, is immaterial. If you grant this you have granted my Axiom II.; for this does not assert any objective fact at all.)

(iv) Do you accept the logical accuracy and permissibility of my remaining definitions and axioms?

(Objections on the score of *convenience* and *simplicity* had better be considered elsewhere.)

(v) Do you admit the formal accuracy of the proofs of propositions in my Books I. and II.?

(N.B. If you admit this there can no longer be any doubt as to the sufficiency of my premises.)

(vi) Do you admit the objective applications of my three Axioms, and therefore of my system of geometry, as discussed in Chap. I. of Part. III.?

(vii) If you admit that there is a theoretical doubt as to the objective counterpart of my second Axiom, please give any criticisms which may occur to you on the remainder of Part III.

Now, if there is no flaw in the line of argument I have adopted, it follows that my conclusions are true, and consequently that any objection taken to them outside this line of argument, however specious it may sound, must contain a fallacy. I might therefore refuse to discuss such an objection. But the objector might truly urge that, conversely, if *his* objection was irrefutable, there must be some hidden fallacy in *my* argument. And therefore, though I prefer arguing in my own way, having devoted a good deal of thought to the subject, and having come to the conclusion that my line of argument is the most direct, and the easiest to discuss; I shall nevertheless feel bound to give the best answer I can to any reasonable objection whatever.

I must here point out that, this book being intended for the study of geometricians, I have not entered upon the question whether beginners could readily be brought to understand it or not. If it is not logically sound, to discuss such a question would be useless. But if it is acknowledged to be logical, I have no doubt that it could be drummed into the head of the average schoolboy as easily as Euclid. But I prefer to postpone this question till the more important one is at least on a fair way towards settlement; when I shall, I hope, bring out a text-book for beginners founded on my method.

EDWARD T. DIXON.

12, BARKSTON MANSIONS,
SOUTH KENSINGTON,
*January* 1891.

# PART I.

## ON THE LOGICAL STATUS OF THE SCIENCE OF GEOMETRY.

## CHAPTER I.

CAN we be absolutely certain of anything in this world? Or is all our knowledge only empirical and approximate? Is there such a thing as *necessary truth*, and if so how are we to know when we have attained it?

These questions open up perhaps the most disputed branches of Logic and Metaphysics. Under one form or another the contest has been raging round them ever since the time of Aristotle. The line of battle has sometimes shifted forward, sometimes back, sometimes it has changed front, so that quite new issues seemed to be at stake. But the status of the Science of Geometry has always been the key of the position; though the combatants on both sides have often confined their energies to flank attacks, in despair of making any impression on the citadel.

It was a prevailing idea among the ancients, though perhaps it was never distinctly formulated, that if a man only had a perfectly clear brain, if, that is, he could only always think logically, he could know everything. Hence the importance which was ascribed to formal logic, and hence the systems of philosophy founded upon data supposed to be known 'a priori.' These fallacious methods would probably have been abandoned much earlier than they were, had it not been for the apparent success obtained in one instance, namely, in

D.                                                                 1

Geometry. Here at least it seemed that a real knowledge of the external world had been deduced from *a priori* considerations alone. When, barely 300 years ago the great Francis Bacon vigorously denounced such *a priori* reasoning, and founded the school of experimental science, neither he nor his disciples appear to have applied their reasonings to upset this accepted view. And that he did not at once revolutionise scientific thought is evident from the celebrated dictum of Descartes, put forward half a century later, that 'everything which we can clearly and distinctly conceive, is true.' Bacon indeed advocated experiment as a means to help the imperfect deductive power of man, rather than as an end. It was left to later philosophers, of whom Mill may be taken as a type, to make a fetish of this means, under the name of Induction, to exalt it above deduction and even to assert that all knowledge whatever is gained by it alone. Such a theory of course involved Mill in an attack on the deductive citadel—the science of Geometry. It is not necessary here to examine the details of this attack (which has moreover been most ably done in a recent work by Prof. Jevons) as the general considerations to be advanced presently will I think be sufficient to repel it. Other philosophers have been driven to take up rather different ground. Kant for example divided all judgements into analytic judgements, in which what is asserted in the predicate is already contained in the connotation of the subject, and synthetic judgements, in which something is added to this connotation. And he maintained that there are certain judgements of the latter class whose truth could be known *a priori*. Among these he classed the Axioms of Geometry, and so cut the Gordian knot. But to further exemplify his views he was ill-advised enough to put the axioms of Arithmetic into the same category, and in this latter case the fallacy is easily exposed, as I hope to show presently.

But let us first consider how we come to know anything at all. Philosophers of the inductive school say 'By induction from experience.' But what is experience? It is easy to put a case on paper—"in so many instances antecedent A was followed by consequent a"—and so on. But how do we know anything about either A or a? A little thought must convince anybody that the only things of which we have any direct cognisance whatever are our own subjective impressions, our sensations and thoughts, and that no one can be *absolutely* certain that such a thing as an objective world exists at all. If I say 'I had a tooth-ache this morning,' I cannot be absolutely certain that the ache was due to a bad tooth, I cannot even be absolutely certain that there is any objective entity corresponding to what I mean by 'my tooth' at all. Besides I may have only dreamed the tooth-ache, or my memory may have played me a trick, perhaps it was not this morning that I felt the ache. But at any rate while the ache is going on I can be *absolutely* certain of its subjective existence; which subjective existence is no less real even if I am dreaming the while. And though in the case of a tooth-ache this truth is painfully evident, it is not less true in the case of the feeblest sensation or thought. We have direct cognisance, each of us, of what goes on in his or her mind, and though we may subsequently forget it, though we may misrepresent it in trying to convey it to the mind of another person, while it is passing in our mind it is to us absolute truth, that is, the thoughts or sensations are absolutely there; though they may, or may not correspond to objective realities in the external world. If therefore we confine Descartes' dictum to purely subjective applications it is strictly true.

Here then we have absolute certainty. Each of us can be absolutely certain of what he is feeling or conceiving at any given moment, and if there are any of his conceptions which he can call up at will he can be absolutely certain that those conceptions have to him a real existence; subjectively, in his own mind, be it understood; though not necessarily objectively, outside it.

In view of these considerations the old classification of branches of knowledge, that is, sciences, as deductive and inductive will need modification. The old designations are indeed very misleading, for there hardly exists a branch of knowledge in which deduction does not play an important part. Perhaps there was little or no true deduction in Astrology; or at least it was an excessive and unwarranted use of induction that led its votaries into error. Probably to the old salt "weather wisdom" is purely inductive. But it is only since deductive methods have been applied to them that Astrology and weather wisdom have entered the ranks of the sciences, as Astronomy and Meteorology. In the same way it was *deduction* not induction that raised Alchemy to Chemistry, and the labour of cataloguing animals and plants to the science of Biology. The real distinction between what used to be classed as 'inductive' and 'deductive' sciences lies, not in their 'methods,' but in the premises upon which they are founded. In the one case the premises are drawn from observation and experiment, and are not only inductive but *objective,* in the other they are directly apprehended by the mind and can therefore only be *subjective.*

It was maintained by some philosophers of the inductive school that inasmuch as the conclusion of a syllogism is contained, or implied, in the premises, no new knowledge can be attained by deduction. Such an argument is however the merest quibble. It is of course true that if you *grant* the truth of the premises you *grant* the conclusion. But to say that therefore you *know* the conclusion because you *know* the premises is a flagrant non sequitur. Why otherwise do we ever learn anything beyond the premises of a science? Can we, because we know the premises of Arithmetic, be said to *know* the product of 3·1416 × 2·7183 before we have multiplied out? It most certainly is possible, very materially, to increase our knowledge by deduction alone; and if there are in our minds any suitable subjective certainties which we can use as premises, it may be possible to establish extensive (subjective) sciences, whose (subjective) conclusions will be necessary truths, as their premises are.

Let us then examine what kinds of premises a deductive science may have. They may be classed under four heads— (1) Objective facts, (2) Postulates, (3) Definitions, (4) Axioms.

Under the head of objective facts I mean to include all facts known by 'observation and experiment,' that is therefore, all objective propositions which we may regard as established truths. I have however already shown that they cannot be *necessary* truths and with them therefore we are not at present concerned.

Postulates are objective propositions about whose truth there yet remains some doubt, or whose truth is only provision-ally asserted until all the consequences that would flow deductively from it have been examined. Or else they are subjective propositions, dealing with an imaginary state of things; and in this case the conclusions drawn from them are their necessary consequences, if the deduction has been strict, though they are not necessarily true. Thus the wave theory of light is founded on the postulate that a certain elastic medium, the æther, pervades all space. This postulate now-a-days almost ranks as an objective fact, for the consequences which flow deductively from it have been found to tally so exactly with what is actually observed. But if it had not been so, the postulate might still have been assumed subjectively, and, as a mental exercise, an imaginary wave-theory of light might have been worked out.

The logical status of a definition has long been a matter of debate. The common view is that, it being a purely verbal proposition, it can convey nothing but philological knowledge. But just as I have shown that knowledge can be truly increased by deduction, so it may be by definition. Thus if I am told that an even number is one which can be divided into two equal (integral) parts, this definition may suggest to me a completely new conception, from which I may be able to deduce a whole new science, which without it I could never have attained. But apart from suggesting something new, if a definition enables us to analyse a concept which we formerly entertained only vaguely, it may enable us to deduce results from it which perhaps we already knew were true, but which we could not before connect with that concept formally. But, lastly, it is a characteristic of that marvellous instrument, Language, that by its aid we are actually able to reason accurately about things we do not clearly conceive; and this we are enabled to do by defining a word or symbol *by its attributes*, without necessarily conceiving its meaning as a whole at all. For example, we may define the symbol $\sqrt{-1}$ in symbolic language thus—

$$\sqrt{-1} \times \sqrt{-1} = -1,$$

and from this definition most important results can be deduced, such as the series from which the values of the trigonometrical ratios are calculated, and this without ever attaching any definite denotation to the symbol $\sqrt{-1}$.

Thus all that is logically required for a definition is one or

more assertions with regard to the word to be defined or its attributes. Having made the assertions it may be that there is no
objective reality corresponding to the word—there may not be
anything in the world possessing the named attribute or combina-
tion of attributes. It may further happen that we know of no
subjective concept corresponding to it; that our minds are not
able to form a concept combining the attributes; but unless the
assertions are inconsistent, that is unless the falsehood of one
can be deduced from the truth of another, the definition remains
logically sound, and we may deduce theorems from it which
may, or may not, turn out useful or interesting, but which at
least are logically true.

In such definitions by assertion it is not necessary that the
word to be defined should be the subject of each proposition, as
is usually the case; but it is necessary that it should be clearly
understood *which* word is being defined. It might indeed be
possible by two assertions each about two unknown words to
define them both, just as by two simultaneous equations we
may define two unknown quantities; but there would seldom
be any advantage in such a course. And in any case the
meanings of all the other words in the definition must be
known. The propounder of a scientific theory is not of course
expected to teach his readers to speak, it is only necessary for
him to define the terms peculiar to his science, or those to
which he wishes to attach peculiar meanings. He may therefore
assume that the meanings of all other words are known to his
readers.

It having been taken for granted that a definition could not
impart any new knowledge, many propositions which are really
definitions have been commonly classed as axioms, especially
where the word to be defined does not appear as the subject of
the sentence, or where it is not quite clear which word requires
definition. Thus Euclid's first eight axioms do not assert anything
which is not contained in the connotation of the words 'equal'
'whole' 'part' &c. So, though they would be very bad definitions
of these words, they are logically nothing more. So also in his
ninth axiom the only question is which of the words is supposed
to be unknown; for this proposition also only expresses what is
part of the connotation of its terms.

A definition, then, lays down the connotation which a word

is to bear, but does not assign to it its denotation; it does not assert that there is any objective thing, or even subjective idea, corresponding to it. It is indeed possible to have a science founded partly, or even entirely, on definitions without asserting that there exist any things corresponding to the terms or symbols defined, and such a science will be ready to hand if any things are found possessing the attributes connoted by the definitions. Algebra is strictly such a science. The symbols used, $a$, $b$, $x$, $y$ &c. are defined, that is, their connotation is laid down, when the distributive commutative laws, and law of indices are asserted about them. But no denotation is laid down for them—it is a mistake to allow beginners to suppose that they are merely numbers in disguise. It is indeed true that, within certain limits, numbers do possess the attributes ascribed to algebraical symbols; and, within those limits, algebraical results are arithmetically true. Within other limits vectors obey the same laws, and hence some algebraical formulæ may be given geometrical interpretations. But pure algebra is independent of such applications, and consists in results deduced from definitions alone, and its logical basis would be just the same even if it had no practical applications.

The last class of premise we have to consider is the Axiom, or necessary truth. Kant defined an 'apodictic' truth as one the negation of which was inconceivable. This involved him in a difficulty, for though you or I may be unable to conceive the negation of a given proposition, perhaps some one else, in a future generation if not now, may be able to do so. Thus any one might assert that "$\sqrt{-1}$ is an impossibility" and might maintain that the assertion was apodictically true—until some-one found a conceivable interpretation for the symbol. We have however found a better criterion of necessary truth above. We have seen that a necessary truth can only refer to the subjective concepts of an individual, which are actually present to his mind, and for an abiding conviction of its truth it is therefore necessary that the concepts be such as his mind can call up at will; and then the realisation of its truth depends upon the direct apprehension of the relations of subject and predicate. For example if I say 'cogito, ergo, sum' I have a perfect subjective consciousness that I do think, and I conclude with absolute certainty that I do (subjectively) exist. But if I were to conclude that my body existed, objectively, I should be guilty of a fallacy. And obviously it would be merely begging the question to say 'cogitat, ergo est.'

The simplest form of axiom, then, is that which merely asserts that a thing, defined in a certain way, is conceivable; or may conceivably undergo certain operations. Every man can pronounce with certainty upon such propositions whether they are true to him or not, if he understands the terms in which they are stated. In the case of those axioms I shall bring forward I cannot doubt that everybody will assent to their truth; but if any exceptional individuals should be found to deny them, that is to deny that they are able to conceive the things named, their denial will only reflect on their individual mental capacities, and need not raise any doubts in the minds of others.

Let us then apply these considerations to the simplest 'deductive' science, that of Arithmetic. Kant maintained that some of the dicta of Arithmetic were 'synthetic judgements *a priori.*' He instanced the equation $7 + 5 = 12$. He was indeed correct in saying that the conception of 12 is not the same thing as that of $(7 + 5)$. But the truth of the above equation can be directly deduced from the conceptions of 7, 5, and 12. This is therefore only another proof that new knowledge may be gained by deduction. For we may define 7 as $1 + 1 + 1 + 1 + 1 + 1 + 1$ and 5 as $1 + 1 + 1 + 1 + 1$, and if these definitions be substituted in the left side of the above equation, it becomes the definition of 12. It is thus evident that as far as simple addition goes, and so also with simple subtraction, Arithmetic is purely a matter of definition. When however we come to multiplication and division we require two axioms which may be stated thus:—

(1) Units may be conceived to be added together to form numbers, which may themselves be treated as units.

(2) Any unit may be conceived to have been formed by the addition of any number of sub-units, which may themselves be treated as units.

Thus, to prove that "$2 \times 3 = 6$" the symbols mean "consider 3's as units, and add two of them together. The result will be the same as six of the original units" which follows from the definitions of 2, 3, and 6. Thus also $3 \times 2 = 6$ and therefore

$2 \times 3 = 3 \times 2$. In this way the commutative law of Arithmetic is established, and the distributive law may be proved equally easily, and so any rule of Arithmetic.

It might indeed be urged that the above two axioms were in reality only two more implicit definitions, that they only helped to define 'unit' and 'number.' That they are, however, something more may be shown by taking an instance in which, though they might still conceivably be true, they might also be false. Thus,—Are two twopenny apples worth fourpence? May we add each of the two pennies for each apple together to form twopences, and then add the twopences as if they were units? The answer is No; there may be a reduction on taking a quantity, or if the vendor wants to eat one of the apples himself he may ask more than fourpence for the two. When he said one apple for twopence he did *not* mean that each twopence might be treated as an unit to make fourpence for the two apples.

We have then established a science whose conclusions, interpreted subjectively are necessary truths. As a matter of fact the premises of Arithmetic are so commonly true among the objective units, as we believe them to exist, that its conclusions are often thought to be as true objectively as they are subjectively. If you grant that there *are* two apples each worth twopence, and that the twopences *may* be added together just as the individual pennies were, then the two apples *are* worth fourpence, but the above objective assumptions are not necessary truths. Besides the perfect subjective Arithmetic we have therefore an objective Arithmetic, whose objective premises are established by induction, and which is therefore imperfect, in the sense of not dealing only with necessary truths.

Well then, in Geometry ought we not to find precisely the same thing? Ought there not to be a perfect subjective Geometry, as well as an applied objective one, the applicability of the former to the latter being a matter to be determined by induction from observation and experiment? It is the object of this book to show that such is the case, to establish the perfect Geometry, and to examine the grounds on which we may believe that it applies to the objective space in which we live.

# CHAPTER II.

I⊤ would be folly to attempt to substitute a new thing for an old one, unless one is first convinced that there was something bad in the old one, which will be remedied in the new. Before therefore attempting to introduce a new system of Geometry I must point out what I consider the defects of the old ones.

That there are some defects has long been recognised. It has for a long while past been pretty generally admitted that Euclid's last Axiom is not a necessary truth. Great is the ingenuity which has been expended by Geometricians in attempting to find an 'apodictic' substitute for this 'Axiom.'

Legendre thought he had proved its truth from the other data of Geometry, but he himself had afterwards to admit that his argument was fallacious. After so conspicuous a failure most Geometricians seem to have reconciled themselves to the idea that the 'Axiom' could never be made apodictic, and to have regarded attempts to do so in the same light as attempts to square the circle. This is, I suppose, the reason that they have overlooked a proof given by M. Vincent of Paris (quoted by Mr F. W. Newman in his "mathematical tracts") which I give, with certain trifling modifications, at the end of this chapter.

But, on the other hand, a great many Geometricians still imagine that the proposition "Two straight lines cannot enclose a space" is an *a priori* necessary truth. If it is only taken to refer to subjective concepts it may indeed be so regarded, if it is taken as the definition of 'straight'; and though it would be by no means a complete definition, as far as it goes it would be irrefutable. But it is perfectly evident that Geometricians do *not* regard it only subjectively. They deduce objective conclusions from it, and it can only be held by them to refer to objective straight lines. Now it is a sufficient definition of an objective straight line to say that two straight lines of a *given small length* cannot enclose a space. From this definition it is possible to mechanically construct 'straight' edges of the given length, on the principle on which Whitworth constructs his surface plates; and two of these 'straight' edges being joined together with only a small part of their respective ends overlapping, we get a longer 'straight' line. By continuing this process it is theoretically possible to produce a 'straight' line to any length whatever, (unless it should at some point rejoin itself.) And how can we be certain that two such 'straight' lines diverging from a given point will never meet again? There is no *a priori* reason why this should be so. All we can say off-hand is that, as a matter of fact 'straight' lines have never been known to behave in such a manner; but we are bound to confess that they have been traced only through an infinitesimal portion of the infinite space we dwell in, if so be that they do not intersect again.

As however this axiom has generally been assumed to be a necessary truth, it has been deduced from it that parallel straight lines, as Euclid defines them, are necessarily possible. For it may be shown that two 'straight' lines which have a common perpendicular must either intersect on both sides of that perpendicular or on neither, since the figure is symmetrical with respect to it. That the reality of parallel lines depends so directly upon this Axiom has not however been generally recognised, simply because Euclid does not prove it in the above way, but by means of his 16th proposition, the proof of which, as it stands, merely begs the question at issue. To complete the proof it is necessary to appeal not only to his 10th Axiom but to another which he does not lay down; to the effect that, in a plane, 'it is impossible to pass from one side of an unterminated line to another, without intersecting it'—an omission which I am glad to see a recent editor[1] has noticed, though he has not noticed the want of the Axiom in this proposition. For in Euc. I. 16, having shown that the triangles *CEF*, *AEB* are con-

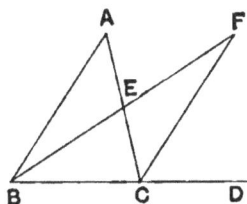

[1] The 'Pitt Press Euclid' by H. M. Taylor.

gruent, and therefore the angles *ECF, EAB* equal, Euclid goes on " But the angle *ECD* is greater than the angle *ECF* (Ax. 9)." Now 'Ax. 9' only says "the whole is greater than its part," and to assume that the angle *ECF* is 'part' of the angle *ECD* is to beg the question at issue. The proof should be completed thus—'Since the line *CEFC* is unterminated, if any part of the straight line *CD* is within *CEEC* it must intersect its boundary twice; for if it is produced far enough there must be points in it in either direction outside this limited space (new Axiom). But *CD* already intersects *PE* in *B*, and *EC* and *FC* in *C*, and so cannot intersect either of them again' (Ax. 10). Therefore *CD* is wholly without the triangle *ECF*, and the angle *ECF* is therefore 'part' of the angle *ECD*, and is less than it (Ax. 9).

Thus, granting neither of the two Axioms, but defining a 'straight' line as one which is *in general* determined if two points on it are given, we have two alternatives to Euclidian Geometry—we might either have more than one 'straight' line through a given point fulfilling Euclid's definition of parallel, with respect to a given straight line, or we might have none at all, in which case any two straight lines which intersected at all would intersect twice. These alternatives have been already examined by Mr G. Chrystal[1] though his treatment of the subject led him to rather erroneous conclusions. For he seems to consider that the former alternative, which he calls hyperbolic Geometry, is the only one worth practical consideration, whereas, as we shall see, it is in reality inconsistent with the self-congruence of space, which he himself assumes, and may be shown to be untenable by the proof given at the end of this chapter of Euclid's Axiom 12.

Let us examine Euclid's premises more closely, in which, as we have seen above, are included his definitions as much, if not more, than his postulates and axioms.

A term generally *denotes* a thing or idea, and *connotes* its attributes. Hence also a definition may indicate the denotation, or detail the connotation of a term.

We have already seen that for deductive purposes the connotation is the only important thing to lay down. Thus Euclid's first definition " A point is that which has no parts or no magnitude"

[1] Paper on 'Non Euclidian Geometry' in 'Proceedings of Royal Society of Edinburgh,' Vol. x., 1879—80.

though it does not tell us what a point *is*, is nevertheless a useful definition, as it tells us at any rate that it does not connote magnitude or the possibility of subdivision. The definition ought not however to have been expressed as if it explained the denotation of the term. An instant of time cannot be said to have parts or magnitude, but 'a point' is not 'an instant.' The definition ought merely to assert, as part of the connotation of the term "A point has no parts and no magnitude." Similarly it is not true that a line *is* length. Length is only one of the attributes of a line. The same remark applies to his definition of a surface. Now how comes it that neither of these two fundamental definitions is ever referred to throughout Euclid's elements? There are I think three reasons. First, Euclid has expressed the most essential part of the connotation of the term 'straight' line elsewhere, namely in his 10th Axiom. Secondly, because he assumes the items of connotation of the term surface, when necessary, without referring to any definition or Axiom. But chiefly it is because, if Euclid ever tried to make any use of these definitions, he would be confronted with the question 'What are length, breadth, and thickness?' A question to which he provides no answer.

Euclid's definition of a 'straight' line is generally acknowledged to be useless. It is merely a paraphrase of the word 'straight,' and by no means a good one. Euclid himself defines a 'plane' surface in a similar manner—the definition substituted for it in modern editions being the only one of the first nine from which any deductions are made, and this one is used illogically, for he never postulates or proves that the thing defined as a 'plane' can exist.

Next we read "A plane angle is the inclination of two lines to one another in a plane, which meet together, but are not in the same direction." What does this mean? If an angle is only another word for an inclination how does this definition assist matters? Or does Euclid recognise the inclination of two lines which do not meet? What has the plane got to do with it? Is the angle between lines which meet but are not in a plane, as between the meridians of longitude at the pole, essentially different from the angle between two intersecting arcs of circles in the same plane? Why does Euclid here admit the word 'direction' which he has so carefully banished from the rest of his work? Why does he trouble to give a special definition of a plane rectilineal angle, unless it is that in this case he can dispense with the awkward word, and pretend that 'straightness' has nothing to do with 'direction,' for which it has been substituted?

Again, why does he speak of angles as *contained* by straight lines? Why did he in I. 16 quoted above, call the angle *ECF part* of the angle *ECD*?

The answers to all these questions are really simple enough, if we only make up our minds to use Euclid's own word, 'direction,' a little more freely. By '*inclination*' Euclid means difference of direction. This is why two curved lines must extend in different directions from their common point, to have an inclination at that point, and it is because he tacitly assumes that a straight line always extends in the same direction from any point in it, that in the case of the definition of a plane rectilineal angle the word direction may be left out. An *angle* is something different from an inclination, which is used to measure it, just as a two-foot rule is not a distance, but is used to measure distances. To Euclid there can be little doubt that an angle meant *a portion of a plane*, a sector cut off between two intersecting straight lines. In the same way a figure, such as a triangle, or circle, was to him not an arrangement of lines, but a piece of a plane; as it were a figure cut out of paper, of which the lines were merely the boundaries. Thus when in I. 4 he says "The whole triangle *ABC* coincides with the whole triangle *DEF*, and is equal to it" he clearly refers to them as triangular pieces of planes whose *areas* are thus proved to be equal. Euclid's angles are really corners of such figures cut out of a plane. This is confirmed by the easy way in which he extends the use of the word angle to the corner of a solid figure where three or more planes meet, calling it a 'solid angle' as opposed to a plane angle, a solid figure being merely *bounded* by the planes, as a plane figure is merely bounded by lines. That this view of an angle is confessedly taken by some Geometricians will appear from M. Vincent's proof quoted below, and from the fact that Mr Newman accepts it without comment. Though perhaps Euclid never acknowledged it to himself, these angles, these sectors of planes, were of course areas, of infinite extent, but such that they bore finite ratios to each other, which ratios measured the inclinations of the bounding lines ; some given sector, or the whole plane, being taken as an unit. Thus if he had ever considered curvilinear angles after defining them, he would either have had to explain that a curvilinear angle was equal to the rectilinear angle between the tangents at the 'angular point,' or he would have become involved in the doctrine of limits.

According to this view then, in the figure of I. 16 the angle *ECF* actually is a *part* of the angle *ECD*. But another common interpretation of the word 'angle' is that it is a quantity of revolution, from one direction to another. This is the way it is regarded in Trigonometry, and it is a way which has many advantages. But when Geometricians introduce it into elementary text books they almost invariably forget that a straight line may be revolved from one direction to another in various ways, describing various conical surfaces the while, and performing various quantities of turning, and that there is therefore no reason why the amount of turning which measures the inclination of *CE* to *CF* should be *part* of that which measures that of *CE* to *CD*, unless the method of rotation is further defined. This is sometimes done by saying the rotation must take place in the plane containing the lines, and that therefore in the above instance the angle *ECF* is only part of the angle *ECD* because *CF* is in the plane of *CE* and *CD*. This would be right enough if the word plane were so defined that we could deduce from the definition that, in revolving in it from *CE* to *CD* a straight line must pass once, and only once through *CF*. But if this result can be obtained otherwise, the conception of a plane may be altogether omitted, and with advantage, as it is by no means a simple concept, and an angle is a sufficiently difficult thing for a beginner to grasp without any unnecessary complications being involved in it.

It is obvious from the above considerations, that an inclination can never be greater than that measured by two right-angles,—or as it is in some modern books conveniently called, a straight angle. This inclination is that between two opposite directions. An Euclidian angle, that is, a sector of a plane, could strictly speaking never be greater than a whole plane, that is, four right angles, unless we are permitted to count some parts of the plane twice. But a Trigonometrical angle, that is, a quantity of revolution (under certain conditions), may be as great as we please, one or more whole revolutions. Thus one inclination may be measured by at least two Euclidian angles, or by any number of Trigonometrical angles. A great deal of the difficulty of Elementary Geometry is really caused by the confusion of these three distinct concepts. My definition of an angle is so framed as to restrict the meaning of the term to the smallest positive trigonometrical angle.

There is little to be said about Euclid's remaining definitions. In that of Parallel straight lines we again come across that curious condition "In the same plane"; a condition which in the subsequent axiom about straight lines which are not parallel, is, curiously enough, omitted from all text books, as far as I know, which gives the axiom at all! The definition of parallel lines is logically good, but of course it remains to be proved that such things really exist. It has already been pointed out how this may be done by the aid of Euclid's 10th Axiom, though Euclid's own method is unsatisfactory.

As to Euclid's postulates, they are logically needed, since definitions cannot be taken to assert the reality of the things they define. But I think it is improbable that it was for this purpose that Euclid wrote them down. To him they merely represented permission to perform certain operations in Geometrical drawing. This is why he has no postulate about drawing planes—he knew of no practical method of drawing them, as straight lines are drawn on the black board, and accordingly he quite overlooks the logical necessity of a postulate or theorem to prove that they are conceivable.

Most of the Axioms have already been discussed. The tenth I have shown not to be a necessary truth, if it is taken objectively, as clearly it is intended to be. The eleventh is quite unnecessary, as it may be proved as a theorem, by the method of superposition. The twelfth however may be proved, if the self-congruence of space is admitted, that is, if we assume that geometrical figures may be carried about in space without any change of size or shape. This proof is virtually that of M. Vincent referred to above.

We have seen already that two 'straight,' that is self-congruent, lines in a plane which have a common perpendicular, must either intersect on both sides of the perpendicular, or on neither. In the latter case, the 'straight' lines are of infinite length, and enclose a part of the infinite plane in which they lie, which we may call a 'band.' If $OA$, $PB$ be two straight lines enclosing half such a band, on one side of the common perpendicular $OP$, and $OP$ be indefinitely produced, to $X$, and $Q$ taken in it so that $QP = PO$, and $QC$ drawn likewise perpendicular to $OP$; $QC$, $PB$ enclose another half-band which may be shown by superposition to be equal to that enclosed by $PB$ and $OA$. And so other bands may be cut off from the plane by straight lines $RD$...&c. But it may be shown by superposition that the remaining quadrant of the infinite plane $XRD$ is still equal to $XOA$. Hence the area of a half band is infinitesimal compared with that of a quadrant of the infinite plane, that is with an Euclidian right angle.

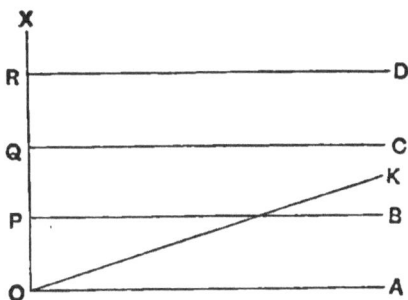

But the area of any finite Euclidian angle is comparable with that of a right angle.

Therefore the area of any finite Euclidian angle is infinitely greater than that of any band of finite width.

Hence if any straight line $OK$ on the same side of $OA$ as $PB$ make any finite angle with $OA$, the whole area of the angle $KOA$ is greater than that of the band $BPOA$ and therefore $PB$ must intersect $OK$.

This proves Euclid's 12th Axiom. The proof does not seem quite satisfactory, chiefly because the accepted definitions of the terms involved do not admit of its being stated in strictly formal language. But as I prove the same thing by a more comprehensive and formal method later on, there need be no doubt as to its correctness.

# CHAPTER III.

But though I have criticised Euclid's premises, and in them those of most systems of geometry, in some detail, I have not yet mentioned what I take to be his fundamental misconception, and that of nearly all subsequent writers on the subject. Even though it is not explicitly stated in Euclid, it seems that he, in common with most modern philosophers, regarded the conception of space as a fundamental attribute of the human mind, which requires no definition or discussion, being necessarily one, and unalterably the same, to every one. This is the view that was taken by Kant, and so energetically combated by Helmholz, and lately again by Mr Herbert Spencer. Helmholz indeed attempted to analyse the conception of space, and got so far as to call it a "Dreifachige Manigfaltigkeit," a three-fold multiplicity, which at least expresses the fact that its constitution is somehow intrinsically connected with the number 3. But Euclid's expression, applied by him to 'a solid,' that is, a limited portion of space, that it has "length, breadth and thickness," really means as much, and might be made to mean a great deal more if length, breadth and thickness were properly defined. As I have already kicked over the traces in making use of the word 'direction,' I may as well say at once that they are measurements made in three different, that is, *completely* different, or as I call them, *independent* directions. If therefore we say 'Space extends from every position in it in three independent directions,' and if further we add 'and the directions in which it extends from any two positions in it are the same,' we have a definition of the concept 'space' in terms of the concept's 'position' and 'direction,' from which it will be seen all the known properties of space can be deduced. And it follows that the conception of space is not a fundamental one at all, but that those of position and direction are more fundamental than it.

It is not necessary, even if it were possible, to explain what the terms 'position' and 'direction' denote. Every one who speaks English must have some idea of their denotation, and for logical purposes it is only necessary to lay down their connotation strictly. Every one knows more or less what is meant by the position of a thing, or the direction of one thing from another, in contradistinction to the thing or things themselves. We may talk of the position of the Solar system, when comparing it with the positions of the fixed stars. But, if we wish to compare the positions of things within a more limited space, it will be evident that there are, within the solar system, many different bodies whose positions differ from each other. Again, within one of these bodies, the earth for example, there are many positions, and if we are dealing with positions within a yet smaller compass, we shall have to use yet more minute bodies to indicate them. Though physicists used to tell us that the 'atoms' of matter were indivisible, the theory has lately been advanced that each atom is a vortex ring of seething æther; and the theory of such vortex rings involves the consideration of differences of position within an atom. When therefore I say that a position may be conceived to be indicated by a minute particle of matter I do not mean to assert any objective truth, but merely to suggest a convenient way of picturing a position to oneself. The second part of the definition of position (for which see Part II.) being added to guard against any confusion of a position with any material thing or point, a position not being a material entity at all.

In the same way a direction does not consist of two points—the points are only used as a convenient way of indicating it. A direction should be conceived as something purely abstract, so that we may speak of two straight lines as extending in identically the *same* direction, not in equal, parallel, or similar directions. This conception of direction has proved a stumbling-block to many—but chiefly, I am convinced, because from their youth up they have been taught to look at geometry from the Euclidian point of view, which attempts only to recognise material dimensions, lengths, areas and Euclidian angles, that is, pieces of planes, and to discard abstract distances and inclinations, that is, differences of position and direction. So much is this idea ingrained in our language and ways of thought, that it is difficult to persuade many, even of those who have never formally learnt Euclid, and much more those who have spent the greater part of their lives in learning and teaching it, that it is possible so to grasp Direction, as an abstract concept. Even when confronted by the enormous development of modern vector theories, even while themselves advocating these theories as one of the greatest developments in modern scientific thought, they will still argue that sameness of direction is, as a matter of fact, an idea deduced from Euclid's definitions of a straight line, and of parallel straight lines; although Euclid himself had spoken of sameness of direction in his eighth definition, prior to giving any workable definition of straightness, or of parallelism. To such objectors it may be replied that it is of very little consequence *how* they obtained the conception of sameness of direction, so long as they have got it; but seeing that, except in the one instance referred to above, Euclid carefully evades the word direction, and that in no text-book or course of oral instruction, so far as I know, is any effort made to explain that 'sameness of direction' means Euclidian parallelism, it is at least very doubtful that they acquired the concept through Euclidian teaching. Nor can it fairly be maintained, until it has been proved by experiment, that those who have not learned Euclid, or become imbued with Euclidian ideas, are incapable of forming this conception. On the contrary, I have tried the following experiment on a few persons who had not studied Euclid—standing a little way from them I extend my arm in any direction, and ask them to extend theirs "in the same direction." Excepting those who were too shy to make the attempt, all have made a more or less intelligent effort to do as I requested them, showing that though their conception of sameness of direction was vague, and probably inaccurate, the conception was there, and only wanted training to develop it. That is to say, they had some idea of its denotation, though the connotation of the term required to be exactly defined to them.

To do this I have laid down four assertions about direction. I have already pointed out that I am logically free to define a word by laying down any assertions I please with respect to it, as long as they are (i) not demonstrably incompatible with each other. But if the definition is to form the basis of a deductive science it is further advisable (ii) that the assertions should be independent, as, if one or more could be deduced from the remainder they might with advantage be omitted, and afterwards proved as theorems. And, where it is required to define logically a term whose denotation is already known, it is further necessary not only that (iii) the assertions should be commonly accepted as true with respect to it, but that (iv) they should restrict the meaning of the term exactly to its accepted denotation, neither more nor less, and should do so in the simplest manner that can be devised. Now my four assertions form a good definition, for (i) they are certainly not incompatible, for all geometricians will admit that (iii) they are as a matter of fact true, about what is commonly understood by the terms direction and sameness of direction, whether they prove this by Euclid or otherwise. The only questions to be examined then are (ii) whether they are independent and (iv) whether they restrict the denotation correctly. I think I may fairly claim to shift the burden of proof of the former proposition on to the shoulders of anyone who denies their independence, merely observing that it has been the attempt to deduce the fourth assertion from the other three which has been the true cause of failure in all previous attempts to make use of 'direction' in elementary geometry. But if anyone should succeed in deducing one of my assertions in the definition of direction from the others, he will in no wise have damaged my theory of geometry, but on the contrary, he will have improved it, by enabling me to substitute a somewhat simpler definition of direction for the one I here present. And as to the fourth question, the sufficiency of the definition, the proof of the pudding will be found in the eating, for no one will say that it restricts the denotation too much, since they have granted clause (iii), and, if the results I deduce are deduced fairly, there can be no doubt as to the sufficiency of the premises.

But though I think these considerations and arguments might be enough to enable my theory of directions to compete with Euclid, if I had a fair start; he unfortunately has the advantage of me by some two thousand years; and I must therefore adduce a few more considerations to show the extreme importance, and fundamental character of the concept Direction.

The idea of a straight line is in common life associated either with the idea of self-congruence, of a line which can twist about itself without apparent motion, of a stretched string, or of a ray of light. From all these conceptions, definitions may be framed, which, as I show in Part III., all lead to the same result. But there remains one conception of a straight line ; that namely which Newton had in his mind when he said that if a material particle were unacted on by any forces, it would remain at rest, or continue to move with uniform velocity in a straight line. There can, I think, be little doubt that the words 'a constant direction' might be substituted for the last three words in this law of motion, without altering the sense. He did *not* mean a straight line as defined by any of the four methods indicated above, for results can be deduced from the law of motion which are not necessarily possible if straight lines are only so defined. From the law of motion it follows that it would be theoretically possible (by gyroscopes, Faucoult's pendula, or other methods) to determine the velocity, and axis, of rotation of the earth *absolutely*, and hence to fix *absolutely* a direction in space, although we cannot fix a position. And if we had two particles the resultant force on each of which vanished, but which remained at the same distance apart, each must be at rest, or moving with uniform velocity in a constant direction. Thus the direction from one to the other is a fixed direction, and since we cannot say they are absolutely at rest, both may be moving in the same direction, though not in the same straight line.

In view of these facts it must be difficult for anyone to maintain that 'direction' is a less fundamental concept than even 'position.' But even if any one should deny the bearing of such a dynamical argument on the elements of geometry, it surely must settle at once the question whether or not a definite meaning can be, or commonly is, attached to the epithet 'same' when applied to directions of straight lines which do not intersect !

The propriety of including the fourth of my assertions in the definition of direction, given in Part II., has been shown logically to follow from the fact that it is acknowledged to be true of what is generally denoted by that word, and that if it were omitted the definition would include more than the word direction is truly held to denote. But it will probably none the less be maintained, by some supporters of Euclidian methods, that the attribute ascribed to direction in this fourth assertion is, to use old fashioned language, an 'accident,' not a 'property' of the term. I have indeed already pointed out that, if so, its truth ought to be deducible from the remainder of the definition; and that if this can be done, so far from upsetting my theory, it would only strengthen it. But I think I can show direct reason for believing that the attribute in question is a *property*; and that as a matter of fact it is always assumed by anybody who uses the expression 'extending in the same directions' in the same sense as 'parallel.'

For surely, if it is conceivable that a point may move in a given direction from $A$ to $B$, or else in two other directions, from $A$ to $C$ and $C$ to $B$ respectively, then must it not follow that if a point may also move from $A$ to $B'$ in the same direction as $A$ to $B$, it may also move from $A$ to some point $C'$ in the direction $AC$ and from $C'$ to $B'$ in the direction $CB$? In books on the theory of vectors, not only is this assumed, but it is further assumed that the transferences in the respective directions must be proportional, or at least, (which comes to the same thing), it is assumed that, in vector language,

$$\alpha + \beta = \beta + \alpha.$$

And so far from attempting to justify this assumption by an appeal to Euclid, it is usual to commence by proving Euclid's proposition that 'the straight lines which join equal and parallel straight lines towards the same parts are themselves equal and parallel,' by it; the above equation being in fact merely Euclid's proposition in vector language. It will be seen therefore that my fourth assertion about direction does not assert as much as the assumption upon which vector theories are based, for it asserts nothing as to the magnitude of the transferences.

Again, if it were required to determine whether two straight lines in space, say two poles stuck in the ground, were 'in the same direction' or not, the first thing that would occur to anyone to do would be to place himself 'in a line' with them, and see whether they 'covered correctly' or not. That is, he would see whether, when one of two intersecting rays of light to his eye (say one along the ground) intersected both poles, and another (say to the top of the far pole, to prevent the chance of its going over the top of either of them) intersected one of the poles, it would also intersect the other. In fact, he would at once apply the test supplied by my fourth assertion about direction. But this assertion has also another aspect. In the former one it corresponds to Euclid's assertion about parallel lines, that they are in one plane. In its other aspect it corresponds to .what has been proposed as a substitute for Euclid's 12th Axiom, being in fact only another way of wording Playfair's Axiom, namely : If a straight line in the same plane with two parallel straight lines intersects one of them, it shall also intersect the other, and this of course is the only aspect of the assertion which is likely to be combated.

But it is here that the difference of the definitions of the terms straight line, parallel, and so on, by direction, or otherwise, comes in. Taking the former system of definitions, but omitting my fourth assertion about direction, if $AB$, $CD$ be the two poles stuck in the ground 'in the same direction,' and $OBD$ be the straight line of sight along the ground intersecting them in $B$ and $D$, the point $B$ being between $O$, the observer's position, and $D$, then as we have seen $OC$ will intersect $AB$, but the question remains, will $OA$ necessarily intersect $DC$ produced ? We can only conceive its not doing so by imagining either $OA$ or $BC$ to be bent outwards—that is, no longer to extend constantly in the same directions. And, as we have not predicated anything concerning the size of the figure, but only concerning its shape (see Part II. def. 15) the difficulty will not be in any way altered on however large a scale the figure be conceived. But if a straight line is defined by any of the other methods, and $CD$ is only asserted to be parallel to $AB$ in the sense that however far they may be produced they do not meet, nothing has been predicated concerning the shape of the figure, and though when drawn of a moderate size its shape

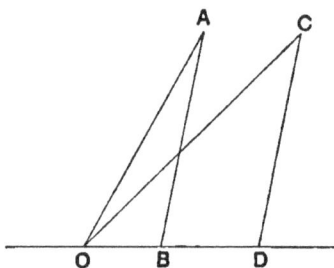

may be, as near as we can tell, the same as if we had used the
definitions by direction, this is no guarantee that the shape
would remain the same if its size were indefinitely increased.
To prove this would require propositions not proved by Euclid
till his sixth book!

(I hope it is unnecessary to remind the reader that the
above appeal to his imagination is not made to prove a
substitute for Euclid's famous axiom, but merely in the hope
of calling his attention to the fact that in reality my fourth
assertion about direction is already part of the connotation he
unconsciously ascribes to that term.)

I have been unable to find in any work on elementary
geometry any formal or satisfactory definitions of the terms
line, surface, and solid, or space. I have already pointed out
that Euclid's definition is of no use, because no definite connotation is ascribed to the terms length, breadth and thickness.
It is however common in more modern text-books to try at
least to describe the denotation of the terms line, surface and
solid in one of two ways. The one way is to begin with a solid,
with which the reader is supposed to be familiar, or with space,
and to say that that which bounds a solid, or separates one
part of space from another, is a surface, and that that which
bounds a surface, or separates one part from another is a line,
and that which bounds a line, or separates one part from
another, is a point. The other way is to begin with a point
and to say that a moving point describes a line, a moving line
a surface, and a moving surface a solid, or space. Often both
methods are given without any effort to show that they lead to
the same results. The first method was formerly, at any rate,
the commonest, but it has one fatal drawback—it assumes that
everybody's conception of a solid, or space, must necessarily be
one and unalterably the same, which I think I can show not to
be the case. The second method pretends to define the conception of space in terms of that of a point, or position ; which
latter conception certainly is of so elementary a character that
we may fairly assume it to be the same to everybody. But in
this method also there is a fatal fallacy—it is, namely, impossible
to define space in terms of a position alone ; direction *must* also
be taken into account. The subterfuges that some geometricians
make use of to disguise this fact are positively amusing. One

of them for example thinks he has done so by making use of the word 'way' instead of 'direction'! Thus, he defines space as "a three-way spread, with points as elements." It is indeed true that a point moving any 'way,' that is, in any direction, from each position in it, describes a line of some sort, and that such a line moving in any way (unless it everywhere moves along itself) describes a surface of some sort; which surface in moving describes a solid of some sort. And if the line is a plane curve and moves in its own plane, it may indeed be said to describe that plane, but parts of the plane may be described twice or three times over, and other parts not at all. And so a curved surface in space, moving in space, may describe part of that space two or three times over and other parts not at all. In order that such a definition of space may be satisfactory and complete, it is necessary to stipulate that the moving point shall move always in the *same* way (that is, direction) or the opposite; that the moving line shall always continue to point the *same* way (that is, that it shall remain parallel to itself), and that a given point in it shall move always in one way or the opposite, different from the ways in which the moving line points. And similar restrictions must be placed upon the movement of the surface so described, in order that it may describe space. These conditions of course involve the conception of direction as I define it, and to evade the use of the word by speaking of 'so many way spreads' is not only begging the question but is positively inaccurate. For, the description of a plane as a two-way spread and a solid as a three-way one is not enough, since a line, even a straight one, has two 'ways,' namely forwards, and backwards, and a plane has not only two, but an infinite number of 'ways.' It is therefore clear that something analogous to my definition of 'independent directions' is necessary to distinguish the three 'ways' which a 'spread' must have to enable it to rank as a solid, besides some workable definition of what a 'way' is. And this definition must clearly contain a clause corresponding to my fourth assertion, for if not the surface, described as above by the motion of a straight line, would not possess the property that any straight line having two points in it would not lie wholly in it; that is, it would not be a plane according to the accepted definition.

# CHAPTER IV.

THERE are a few more points among my premises to which I must briefly call attention. The conception of 'opposite' directions follows as an obvious consequence of the fact that *two* points are required to indicate a direction, and that therefore when we name one, we at the same time indicate a second direction, which stands in a peculiarly intimate relation to the first. From the definition of dependence of directions we see that a direction is 'dependent' upon its opposite direction, but is independent of any other single direction; for the movements contemplated in that definition are merely movements in certain straight lines, and may be either one way or the other along them. (N.B. there is no assumption, or assertion about this—it is merely a matter of verbal definition.) This definition of independence of direction may at first seem wordy, and arbitrary; but the reader will soon see that it merely expresses in precise language what we meant (nine) pages back when we agreed that space might be said to extend in three 'independent' directions from any given position in it.

I draw a distinction between distance and a linear dimension, such as length, which I think will be found convenient. A distance is the difference[1] between two positions, and is an

---

[1] I feel bound to protest here against a criticism of the late Prof. De Morgan in a review of a book of Mr Wilson's, which he published in the *Athenæum*, and to express my surprise that so accurate a thinker could have been guilty of it. He says à *propos* of a definition of Mr Wilson's, "Is a direction a magnitude? Is one direction greater than another? We should suppose so, for an angle, a magnitude to be halved and quartered, is the 'difference of direction' of 'two straight lines which meet one another.'" Clearly neither Mr Wilson nor I use the word difference in the sense it is used in Arithmetic, when we say that 5 is the difference of 12 and 7! If two positions or directions are not the same, they must be different, and that difference is called a distance, or inclination, as the case may be. These 'differences' are not magnitudes, but the magnitudes used to compare them are lengths, and angles, respectively. If Mr Wilson did not make this clear, I hope at least that I have done so.

abstract notion—it does not necessarily imply that there is, or even might be, a straight line between them. Thus it may be measured by an amount of transference, under fixed conditions. But length is a concrete measure of an actual figure; it is a piece of a straight line of given dimension, which may be carried about and compared with other straight lines. So other dimensions may be regarded as material things, pieces of planes or solids. On this view also an inclination is not a dimension but an abstract notion, analogous to a distance; but an Euclidian angle, a piece of a plane, even though an infinite piece, is precisely analogous to an area. These conceptions are brought out clearly enough in the definitions.

The assigned methods of measuring distances and inclinations are purely arbitrary, and that they are feasible has to be deduced from the axioms, later on. It is also shown in the text that the methods are convenient, as they measure the quantities with the minimum amount of transference and twisting respectively. The method of measuring a straight angle is an obvious extension of that used for any other angle; but, in as far as it is arbitrary, it too is shown to be feasible and convenient.

One of the features of my method is that I do not separate the text in the usual manner into 'Plane' and 'Solid Geometry,' but rather into the geometry of lines, and the geometry of surfaces and space. In ordinary Plane Geometry all the figures are supposed to be 'in a plane.' This is rather hard on a beginner, who can hardly be expected to have any clear idea of what a plane is, and the result is that he is apt to miss the force of this important restriction. Even his teachers do the same thing, as witness the fact already referred to that in, I believe, no text-book is the necessary condition noted that the straight lines in Euclid's 12th Axiom must be 'in one plane.' Again in Euc. I. 4, we read, "Therefore the whole triangle *ABC* coincides with the whole triangle *DEF*". I am sure that nine people out of ten miss the meaning of this sentence, which is, that their *planes* coincide. For why should they do so? Did Euclid ever lay down an axiom that two planes cannot enclose a space? And if this fact is deducible from his other axioms, where is the theorem in which he deduces it? Yet this little sentence, which I am convinced most people overlook as a piece of Euclidian circumlocution, is the basis of a very large portion of Euclid's reasoning, including his celebrated 47th proposition, and the greater part of his second and sixth books!

In my method I postpone all mention of planes, even their definition, till the second book. In the first book, however, I prove everything Euclid proves in his first book, which does not refer to plane areas; and two propositions which Euclid gives (though in slightly different forms) in his eleventh book. I am thus able to postpone the definitions of the terms surface, plane, &c., to the second book; where also I give definitions of terms which are either not defined at all in ordinary text-books, or whose definitions would require modification to be brought into accordance with my system. In my second book I prove everything in Euclid's eleventh (except one proposition referring to proportion) and several other highly important propositions which do not appear in any of the ordinary text-books.

My three axioms, being the foundations of a purely subjective science, merely assert the power of the human mind to conceive certain things, the nature of the things in question being explained in the definitions. They do not assert any objective fact at all. Of course it would be possible for any given man to deny that he could conceive those things. I do not anticipate coming across many such people, but as they would be no more capable of understanding Euclid than of understanding my geometry, I do not at present care to argue with them.

My second axiom might be paraphrased thus—Space may be conceived to extend from every position in it in the same directions. For if the positions can be conceived, of course points may be conceived to occupy them, forming straight lines &c. The full import of this axiom can only be explained later on, but in case anyone should object that Euclid *proves* that a straight line can be drawn through any point parallel to any given straight line, whereas I have to assume it, I must refer him to my remarks above, where I show that Euclid's pretended proof really rests on the debatable assumption contained in his 10th Axiom.

I have deferred stating my third axiom till the beginning of the second book. I do this, not only because the beginner would not easily understand it until he had become more familiar with the idea of independent directions, but also to emphasise the fact that the propositions in my first book in no way depend upon it.

As the system of geometry I present in Part II. is a purely subjective one, I do not of course require any powers to *draw* straight lines, or perform any constructions as Euclid does. I only ask the reader to conceive the lines and figures, not to draw them, and only add the diagrams in the book to aid his imagination—he is on no account to suppose that the words in the text refer to the actual lines in the diagrams. Euclid himself virtually abandons his geometrical drawing in his eleventh book, and falls back on hypothetical constructions. But on the other hand *my* imaginary constructions are hardly hypothetical. Each of them is strictly justified by appeals to the axioms of the science. Thus, I *prove* by those axioms that an angle may be conceived to be measured by the method I lay down in the definition,—that a plane may be conceived through any given position extending in any two given independent directions,—that a plane may be conceived to revolve about a fixed straight line in it, and so on. These, and similar points, are overlooked in every other geometry with which I am acquainted, even where actual constructions are made, as in Euclid.

I have presented Part II. of this book almost in the form I should propose to give to an elementary text-book of geometry, though perhaps a few of these explanations would have to be incorporated with it. A beginner might very well assume its objective application at first, it would not be necessary or advisable to trouble him with such metaphysical subtleties. It would evidently be easy upon this foundation to build up the remainder of a complete work of Elementary Geometry. But in the meanwhile, all the fundamental propositions having been demonstrated (including the fact that two planes cannot enclose a space, required in Euc. I. 4), the superstructure may be bodily taken from the ordinary text-books.

### NOTE.

I feel it necessary here to refer briefly to some previous attempts to make use of 'direction' in elementary treatises on geometry, and to the strictures which have been passed upon them. For I have already found a disposition on the part of my opponents in argument to save themselves trouble by shifting the above-mentioned criticisms bodily on to my shoulders. Now I am of opinion that the reviews to which I specially refer, where they resist the temptation to be funny, are in the main just and logical. Neither Mr Wilson, nor Mr Willock, the chief exponents of ' the direction theory' make any attempt to give a workable definition of ' direction,' or 'sameness of direction ;' and their definitions of ' angle' are hopeless jumbles of the three concepts I have taken pains to distinguish above. The consequence is that they lay themselves open to the accusation of having defined parallel straight lines as such as make equal angles with any transversal. Such a charge could not lie against me, for I on the contrary, define sameness of direction, that is parallelism, first, and 'angle' by it. Besides this, not having defined 'sameness of direction' properly they either assume Euclid's 12th Axiom after all, or are led into positive fallacies in the attempt to do without it.

But in any case I think I may claim from my critics the courtesy of first-hand criticism, and that they shall not dish up against me the stuff that somebody else wrote, against some other theory, at some other time.

# PART II.

A SUBJECTIVE THEORY OF GEOMETRY, DEDUCED FROM THE TWO FUNDAMENTAL CONCEPTS, POSITION, AND DIRECTION.

## BOOK I.

### ON STRAIGHT LINES, AND ANGLES.

### DEFINITIONS.

1. IMPLICIT definition of Position:—

(a) A position may be conceived to be indicated by a portion of matter, called a point, which is so small that for the purpose in hand variations of position within it may be neglected.

(b) But a position is not the same thing as a point, for a point may be conceived to move, that is, to change its position, whereas to talk of a position as moving, is a contradiction in terms.

2. Implicit definition of Direction:

(a) A direction may be conceived to be indicated by naming two points, as the direction from one to the other.

(b) If a point move from a given position constantly in a given direction, there is only one path, or series of positions along which it can pass. (Such a path may be called a 'direct path,' and a continuous series of points occupying such positions, a straight line; see definition 4.)

(c) If the direction from $A$ to $B$ is the same as the direction from $B$ to $C$, that from $A$ to $C$ is also that same direction.

(*d*) If two unterminated straight lines which intersect, are each intersected by a third straight line in two separate points, any unterminated straight line extending in the same direction as this last one, which intersects one of the two former, shall also intersect the other.

3. The direction from $B$ to $A$ is said to be *opposite* to that from $A$ to $B$.

4. A *straight line* is a continuous series of points extending from each of them in the same two opposite directions.

(NOTE. Since a straight line which extends from $A$ to $B$ also extends from $B$ to $A$, it must always extend in two opposite directions.)

5. If a point may be conceived to move from one position to another in a given direction, or else along a series of straight lines extending in a succession of directions, the single direction is said to be *dependent* upon those other directions. But if it is impossible to move the point from the one position to the other by any series of straight lines whatever, in those directions, the single direction is said to be *independent* of them.

6. The difference between two positions is called the *distance* between them.

It is conventionally measured by the amount of transference required to move a point from the one position to the other in a constant direction, that is, along a straight line. (N.B. The propriety of this convention will appear when it is shown, in Prop. 16, that the amount of transference so required is a minimum.)

7. The difference between two directions is called their *inclination* to one another.

The measure of an inclination is called an *angle* and is the amount of twisting required to turn a straight line from the one direction to the other, with the condition that in turning it shall pass through a continuous series of directions all intermediate between and dependent upon the directions whose inclination is to be measured. (N.B. The possibility of so measuring an inclination is shown in Prop. 5, and its propriety appears when it is shown in Prop. 17 that the amount of twisting so required is a minimum.)

8.  The measure of the inclination of two opposite directions is called a *straight angle.*

A straight angle is measured by the twisting of a straight line from the one direction to the other keeping it always in directions dependent upon the opposite directions and some third direction chosen arbitrarily. (N.B. Since no third direction can be dependent upon two opposite directions, the ordinary convention fails in measuring a straight angle. The propriety of this supplementary convention appears however when it is shown in Prop. 6, that whatever third arbitrary direction is chosen the result will be the same.)

9.  An angle equal to half a straight angle is called a *right angle.*

10.  The phrase '*parallel to*' may be taken to be equivalent to ' extending in all the same directions as...'

Hence two straight lines are said to be parallel, if they both extend in the same two opposite directions.

11.  The phrase '*perpendicular to...*' may be taken to be equivalent to ' equally inclined to all the directions in which... extends.'

12.  A *triangle* is the figure formed by the three straight lines joining three points, which are not in one straight line, two and two.  The points are called the *corners* of the triangle.  The angle between the directions from one corner to the two others is called an *interior* angle (or merely an angle) of the triangle, and that between the direction from one corner to a second, and that from the second to the third is called an *exterior* angle.

13.  A *quadrilateral* consists of four straight lines joining four points, no three of which are in a straight line, two and two, in order.  If the sides form two pairs of parallel straight lines, it is called a *parallelogram* (for names applied to other quadrilaterals, see end of Prop. 23 of the first book.)

14.  If two figures can be conceived, (either at once or at different times,) to occupy exactly all the same positions, they are said to be *congruent* to one another.

15.  The *size* of a geometrical figure is determined by the distances between the various points in it; its *shape* by the inclinations of the directions from one to another of the various pairs of points in it.

## Axioms.

I. Any geometrical figure may be conceived to be moved from any one part of space to any other, except in so far as it is restricted by the other axioms (see Book II. Prop. 14, corollary (iii)), without its size or shape being in any way altered.

II. A straight line may be conceived to extend from any given position to any distance, in any given direction.

(N.B.   A third axiom is added at the beginning of Book II.)

## NOTE.

A direction may be 'given' by naming two points, as the direction from one to the other, (D. 2. *a.*) or if it is not important to distinguish it from the opposite direction, it is enough to name a straight line extending in it. Hence by the second axiom we may conceive a straight line through one given point extending to another, or, as it is called, 'joining' the two points; we may conceive a terminated straight line to be produced either way as far as we like; we may conceive a straight line through one given point, extending in the direction from a second given point to a third; or we may conceive a straight line through a given point parallel (see D. 10.) to a given straight line.

The expression 'Join $AB$' is used as an abbreviation for 'Conceive a straight line joining the points $A$ and $B$.'

A position, or point, is denoted by a single letter of the alphabet, as 'the point $A$.'

The distance between two positions, or the terminated straight line between them when considered in reference to its length, is denoted by the two letters denoting the positions or extremities of the line, with a bar over them thus, 'the distance $\overline{AB}$,' or 'the straight line $\overline{CD}$.'

A direction is denoted by the letters denoting the points which indicate it, *taken in order*, as 'the direction $AB$.' Thus the opposite direction to $AB$ is 'the direction $BA$.'

An angle may be denoted as 'the angle between the directions $AB$, and $CD$.' But if it is between the directions from one point to two others it is merely denoted by the three letters denoting the points, that denoting the point from which the directions lie being placed in the middle. Thus the angle between the directions $AB$ and $AC$ is called simply 'the angle $BAC$' or 'the angle $CAB$.' A straight line may be denoted by the letters denoting two or more points in it. If there are only two letters the order is indifferent, but if there are more than two they are (if possible) placed in the order they actually lie, either forwards or backwards, but this rule need not be followed where the order of the points themselves is not determinate.

N.B. It must be observed that a direction (and consequently an inclination and an angle also) has no position in space. Thus we may freely talk of the angle between the directions *AB* and *CD* even if the straight lines *AB* and *CD* do not intersect, and to speak of the angle between *AB* and *AC* as 'the angle at *A* ' (as Euclid sometimes does) is incorrect, as also it is to talk of 'the angle between two straight lines.' Moreover as each straight line has two opposite directions, there are four angles 'between them,' not one, if we consider the phrase merely an abbreviation for 'between their directions.'

## PROPOSITION I.

*Two parallel straight lines cannot intersect, nor can any two straight lines intersect in more than one point.*

For, if two parallel straight lines had a common position, and from this position a point were to move in either of the two opposite directions in which they both extend from it, it could only move along one path [D 2 (*b*)] and therefore there could not be more than one straight line through this position in the given directions.

And if any straight line extend from a point *A* in a given direction to the next point *B*, and from *B* to the next point *C*, the direction *BC* is the same as *AB* [D 4], and therefore the direction *AC* is the same direction [D 2 (*c*)]. So it may be proved that the direction *AD* to the next point, and so to any point on the same side of *A* is the direction *AB*. In the same

way the direction from *A* to any point on the other side of *A* is *BA*.

Therefore no two straight lines can have two common points, for if they had they would extend in the same directions, and also intersect, which is impossible (see above).

Therefore *Two parallel straight lines......&c.*        Q. E. D.

Corollary. *Two straight lines cannot enclose a space.*

NOTE. The references in brackets refer to the premises or previous propositions : D=definition, A, axiom, P, postulate, c, corollary.

## PROPOSITION II.

*If each of two unterminated straight lines intersect each of two intersecting straight lines in two separate points, they shall either be parallel to one another, or shall intersect.*

For if two straight lines $AB$, $A'B'$ each intersect each of the two intersecting straight lines $AA'$, $BB'$, which intersect in $O$, in the separate points $A$, $B$ and $A'$, $B'$,

Conceive a straight line through $A'$ parallel to $AB$.　　[A 2.

Then the straight lines $OAA'$, $OB$ which intersect in $O$, are

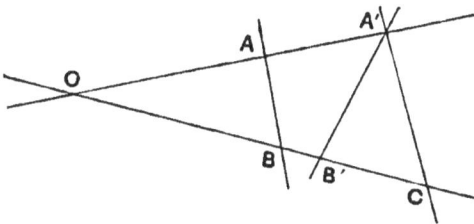

met by the straight line $AB$ in separate points, and one of them, $OAA'$ is met by the parallel straight line $A'C$ in $A'$.

Therefore $OB$ is also met by this straight line in some point $C$ [D 2 (*d*)]. Now, if $C$ is the same point as $B'$, the straight line $A'C$ is the same as $A'B'$, [I. 1], which latter is therefore parallel to $AB$.

But if $C$ is not the same point as $B'$, then the two straight lines $CB'B$, $A'B'$, which intersect in $B'$, are met by the straight line $A'C$ in separate points, and one of them, $CBB'$ is met by the parallel straight line $AB$ in $B$.

Therefore the unterminated straight lines $AB$, $A'B'$ must intersect.　　　　　　　　　　　　　　　　[D 2 (*d*).

Therefore *If each of two......&c.*　　　　　　Q. E. D.

## PROPOSITION III.

*If each of two unterminated straight lines intersect each of two parallel straight lines, they shall either be parallel to one another, or shall intersect.*

For if two unterminated straight lines $AB$, $A'B'$ each

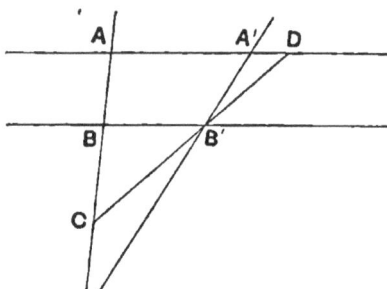

intersect each of two parallel straight lines $AA'$, $BB'$, in $A$, $B$, and $A'$, $B'$,

Let $C$ be any third point in $AB$. Join $CB'$.      [A 2.

Then the straight lines $CBA$, $CB'$ which intersect in $C$, are met by the straight line $BB'$ in separate points, and one of them, $CBA$ is met by the parallel straight line $AA'$.

Therefore the other, $CB'$, also intersects $AA'$ in some point $D$.      [D 2 (d).

Now, if $A'$ is the same point as $D$, the straight line $A'B'$ is the same as the straight line $CB'D$, and therefore intersects $AB$ in $C$,      [I. 1.

But if $A'$ is not the same point as $D$, the unterminated straight lines $AB$, $A'B'$ each intersect each of the two intersecting straight lines $AA'D$ and $CB'D$ in separate points.

Therefore they are either parallel, or intersect.      [I. 2.

Therefore *If each of two......&c.*      Q. E. D.

## Proposition IV.

*If each of two straight lines which intersect are intersected by a transverse straight line in separate points, any straight line through their point of intersection, which also intersects the transverse straight line, is in a direction dependent upon their directions.*

*And conversely, any straight line through their point of intersection in a direction dependent on their directions either intersects, or is parallel to, any such transverse straight line.*

For if $OA$, $OB$ be two straight lines intersecting in $O$, and

*AB* be a transverse straight line intersecting *OA* in *A*, and *OB* in *B*,

And if *OX* be any other straight line through *O*, intersecting *AB* in *X*,

Through *X* conceive an unterminated straight line *XZ*, parallel to *OA*. [A 2.

Then, since the straight lines *BXA*, *BO*, which intersect in

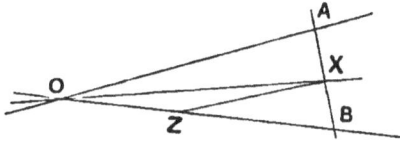

*B*, are met by the straight line *AO* in separate points, and one of them *BXA* is met by the parallel straight line *XZ*, in *X*,

Therefore *BO* is also met by *XZ* in some point *Z* [D 2 (*d*)]. Hence a point may be conceived to move from *O* to *X* in the direction *OX* along the straight line *OX*, or along the straight lines *OZ*, *ZX*, in the directions *OB*, *OA*.

Therefore the direction *OX* is dependent upon the directions *OA*, *OB*. [D 5.

And conversely,

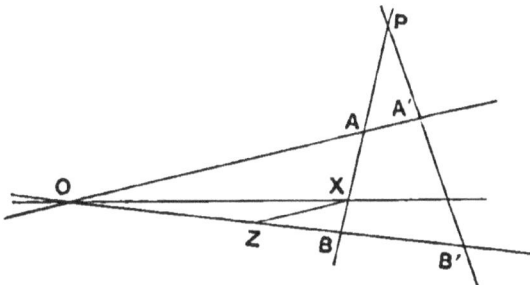

If the direction *OX* is dependent upon the directions *OA*, *OB*,

Then it must be possible to move a point from *O* to some point *X* in *OX* in the direction *OX*, or else to move it to *X* by a series of straight lines *OZ*, *ZX* in the directions *OB*, *OA*.

[D 5.

Let *B* be any other point in *OZB*, join *BX*.

Then since the straight lines *BZO*, *BX*, which intersect in *B* are met by the straight line *ZX* in separate points; and one of them *BZO* is met by the parallel straight line *OA* in *O*,

Therefore $BX$ is also met by $OA$ in some point $A$ [D 2 (*d*)] and therefore $OX$ intersects the transverse straight line $AB$.

And if $A'B'$ be any other transverse straight line, intersecting $OA$ in $A'$, and $OB$ in $B'$,

Then the two straight lines $AB$, $A'B'$ each intersect each of the two intersecting straight lines $OAA'$, $OBB'$ in separate points.

They are therefore either parallel, or they intersect.    [I. 2.

But if they are parallel,

Then the straight lines $OBB'$, $OX$ which intersect in $O$, are met by the straight line $BX$ in separate points, and one of them, $OBB'$, is met by the parallel straight line $B'A'$ in $B'$.

Therefore $OX$ also intersects $B'A'$.    [D 2 (*d*).

And if $AB$, $A'B'$ are not parallel, let them intersect in $P$.

Therefore the straight lines $PA'$, $OX$ each intersect each of the intersecting straight lines $PAX$, $A'AO$ in separate points.

Therefore $PA'B'$ and $OX$ are either parallel, or they intersect.    [I. 2.

Therefore *If each of two......&c.*    Q. E. D.

## PROPOSITION V.

*A straight line may be conceived to be twisted from any one given direction to any other which is independent of it, so as to measure the angle between them, in one, and only one way.*

For if $O$ be any fixed point, straight lines $OA$, $OB$ may be conceived in the given directions from $O$ to points $A$, $B$ [A 2], which will not be one and the same straight line, since the directions are independent.

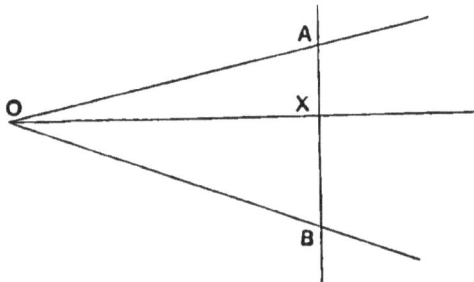

And a straight line may be conceived joining $AB$, and also one from $O$ to any point $X$ in $AB$.    [A 2.

And the point $X$ may be conceived to move along $AB$ from $A$ to $B$, that is in the constant direction $AB$, in one and only one way. [D 2 (*b*).

As it does so $OX$ is twisted from the direction $OA$ to $OB$ in one, and only one way.

And since the directions in which $OX$ extends are always intermediate between, and dependent upon the directions $OA$ and $OB$ [I. 4], the twisting of $OX$ measures the angle $AOB$. [D 7.

Therefore *A straight line*......&c. Q. E. D.

Corollary. *If any direction be intermediate between and dependent upon two others, the sum of the angles it makes with those two directions is equal to the angle they make with each other.* For the twistings of a straight line to measure the former angles are parts of the twisting which measures the latter.

PROPOSITION VI.

*The inclination of two opposite directions is a constant inclination; and the conventional method of measuring it always gives the same angle.*

For if $OA$, $OB$ be straight lines from any point $O$ to points $A$, $B$, in opposite directions,

And if $O'A'$, $O'B'$ be two other straight lines from any point $O'$ to points $A'$, $B'$, in any other two opposite directions,

Then since the direction $AO$ is opposite to $OA$ [D 3], it is the same as $OB$. Therefore $AOB$ is one straight line.

So also $A'O'B'$ is one straight line.

Therefore if the straight line $AOB$ be conceived to be moved so that $O$ falls on $O'$ and $A$ in $A'B'$ [A 1], the straight lines will become one and the same straight line [I. 1], that is $OB$, $O'B'$ will also coincide.

Therefore the inclination of $OA$ to $OB$ is the same as that if $O'A'$ to $O'B'$ in the new position, and the inclination was not altered by the movement of $AOB$ [A 1]. Therefore the inclinations of the opposite directions were the same.

And if the inclination of $OA$ to $OB$ be measured by twisting a straight line from $OA$ to $OB$, keeping it in directions dependent on the directions $OA$, $OB$, $OC$, [D 8.

And if the inclination of $O'A'$ to $O'B'$ be measured by a

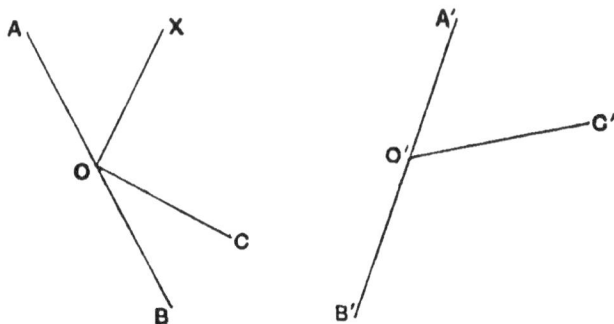

straight line keeping it in directions dependent on $O'A'$, $O'B$, $O'C'$; $OC$, $O'C'$ being any arbitrarily chosen directions,　　[D 8.

Then either the angles $AOC$, $A'O'C'$ must be equal, or one must be greater than the other, say $AOC$ than $A'O'C'$.

In the latter case as a straight line $OX$ revolves to measure the angle $AOB$, before reaching $OC$ it will reach some situation $OX$ such that the angle $AOX$ is equal to $A'O'C'$.

Hence if the whole figure $ACB$ be conceived to be moved and placed so that $O$ coincides with $O'$ [A 1], $AOB$ with $A'O'B'$, either $OC$, or in the latter case $OX$, may be made to coincide with $O'C'$.

Therefore in the former case the figures are congruent, and therefore the twisting of the same straight line measures both straight angles, which are therefore equal.

In the latter case the straight angle $AOB$ measured by a straight line twisted through directions dependent on $OA$, $OB$, and $OX$ is equal to the straight angle $A'O'B'$, and since this straight line in revolving passes through $OC$ [I. 4], it gives the same angle as one twisted through directions dependent on $OA$, $OB$ and $OC$.

Therefore *The inclination......&c.*　　　　Q. E. D.

Corollary. (i) *Hence the sum of the angles between any direction and two opposite directions is a straight angle.*

(ii) *Hence also a right angle, which is half a straight angle, is a constant angle.*

## PROPOSITION VII.

*The angle between two directions is equal to that between the directions opposite to them.*

For if *AOA'*, *BOB'*, be any two straight lines through any point *O*; *OA*, *OB* being two given directions, and *OA' OB'* the directions opposite to them,

If we conceive one end, *OX* of a straight line *XOX'* to revolve from *OA* to *OB*, measuring the angle *AOB*, the other end *OX'* will at the same time be revolving from *OA'* to *OB'*

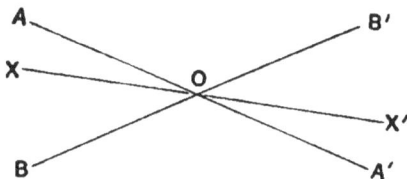

and measuring the angle *A'OB'*.

Therefore the angle between *OA* and *OB* is equal to that between *OA'* and *OB'*.

Therefore *The angle......*&c.      Q. E. D.


## PROPOSITION VIII.

*The sum of the interior angles of any triangle is a straight angle.*

For if *ABC* be any triangle,

Conceive the side *BC* to be produced to *D*,      [A 2.

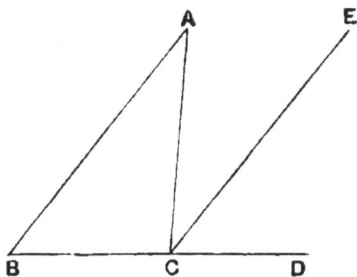

And through *C* conceive a straight line *CE* in the direction *BA*, to any point *E*.

Then because the directions *CE*, *CD* are respectively the same directions as *BA*, *BC*,

The angle *DCE* is the same as the angle *CBA*.

And because the directions *CE*, *CA* are respectively opposite to the directions *AB*, *AC*,

The angle *ECA* is equal to the angle *BAC*.                    [I. 7.

But since a point can be conceived to move from *B* to *A* in the direction *CE*, or else by straight lines *BC*, *CA* in the directions *CD*, *CA*,

The direction *CE* [D 5] is dependent on the directions *CD*, *CA* ; and it is intermediate between them.

Therefore the sum of the angles *DCE*, *ECA* is equal to the angle *DCA*.                                          [I. 5 c.

Therefore the sum of the angles *CBA*, *BAC*, *ACB* is equal to the sum of the angles *DCA*, *ACB*.

That is, to a straight angle.                         [I. 6 c. (i).

Therefore *The sum......&c.*                              Q. E. D.

Corollaries.  (i) *Hence the exterior angle of a triangle, as* ACD *is equal to the sum of the two interior opposite angles* CBA, BAC.

(ii) *Hence also the sum of any two angles of a triangle is less than a straight angle, and any exterior angle is greater than either of the interior opposite angles.*

### PROPOSITION IX.

*If each of two straight lines intersect each of two intersecting straight lines, or of two parallel straight lines, in separate points, and*

(i)  *if their directions on the same side of one of the transverse straight lines are equally inclined to one of its directions;* or

(ii)  *if they make with its opposite directions angles whose sum is a straight angle; or*

(iii)  *if their directions on opposite sides of the transverse line make equal angles with its opposite directions;*
  *the straight lines shall be parallel.*

For if each of two straight lines *AB*, *CD* intersect each of the two straight lines *EF*, *KL*, which either intersect or are parallel,

Then *AB, CD* either intersect or are parallel.　　[I. 1 and 2.

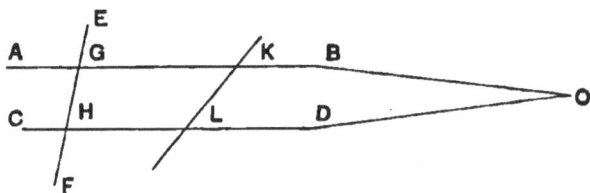

But they cannot intersect, for if they did, in *O* say:

(Then if *EF* intersect *AB* in *G* and *CD* in *H, OGH* would be a triangle, and therefore)

(i)　The exterior angle *EGO* would be greater than the interior opposite angle *GHO* [I. 8 *c.* (ii)], whereas it is not, if the directions *GB, HD* of the straight lines *AB, CD* on the same side of *EF* are equally inclined to the direction *HE*, or *GE*, of *EF*;

(ii)　The sum of the interior angles *BGH, DHG* of the triangle *OHG* would be less than a straight angle.　　[I. 8 *c.* (ii).

And since the sum of the four angles *AGH, BGH, CHG, DHG* is two straight angles　　　　　　　　　　　　[I. 6 *c.* (i).

The sum of the angles *AGH, CHG* would be greater than a straight angle,

Whereas in both cases they are equal to a straight angle, if the directions of *AB* and *CD* on the same side of *EF* make with its opposite directions angles whose sum is a straight angle,

(iii)　And the angle *AGH* would be greater than the interior opposite angle *DHG* of the triangle *OHG* [I. 8 *c.* (ii)], whereas it is not, if the directions *GA, HD* of the straight lines *AB, CD* on opposite sides of *EF* make equal angles with its opposite directions.

So it may be proved that none of the conditions in the enunciation can hold unless *AB* is parallel to *CD*.

Therefore *If each of two......*&c.　　　　　　　Q. E. D.

Corollary.　*And, conversely, it is evident that if* AB *is parallel to* CD *all the conditions in the enunciation will hold*, from the definitions of the terms, or as special cases of Propositions 6 cor. (i), and 7.

## Proposition X.

*If two triangles have two sides of the one respectively equal to two sides of the other, and the angles between the directions of those sides equal, they shall be congruent to each other.*

For if the sides $\overline{AB}$, $\overline{AC}$ of a triangle $ABC$ are respectively equal to the sides $\overline{DE}$, $\overline{DF}$ of a triangle $DEF$,

And the angles $BAC$, $EDF$ between the directions of these sides are equal,

Then the triangle $ABC$ may be conceived to be moved,

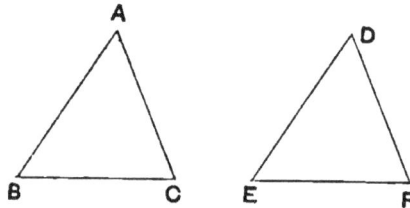

[A 1], and placed upon $DEF$, so that the corner $A$ falls upon $D$, the side $AB$ upon $DE$ and, since the angles $BAC$, $EDF$ are equal, the side $AC$ upon the side $DF$.

And since $\overline{AB}$ is equal to $\overline{DE}$ and $\overline{AC}$ to $\overline{DF}$, the corner $B$ will fall upon $E$, and $C$ on $F$.

Therefore $\overline{BC}$ will coincide with $\overline{EF}$ [I. 1] and consequently the triangles are congruent. [D 14.

Therefore, *If two triangles......&c.* Q. E. D.

Corollary. *Hence if two sides of a triangle are equal, two of its angles, namely, those between the directions of the equal sides and the third side, are equal.*

For if $\overline{AB}$ in the above proof had been equal to $\overline{AC}$, the triangle $ABC$ might also have been moved so that $\overline{AB}$ fell on $DF$, and $\overline{AC}$ on $\overline{DE}$, and the triangles would also have been congruent so. Hence both the angles $ABC$ and $ACB$ would be shown to be equal to $DEF$, and therefore to each other.

## Proposition XI.

*If two triangles have two angles of the one respectively equal to two angles of the other, and have two corresponding sides equal, they shall be congruent to each other.*

For if $ABC$, $DEF$ be two triangles which have two angles of the one respectively equal to two angles of the other,

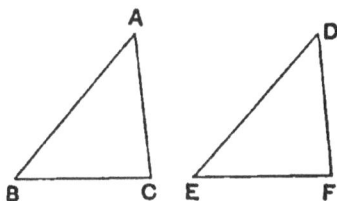

Then since the sum of the three angles of every triangle is a straight angle, [I. 8.

The third angles of the two triangles are also equal.

Hence if $\overline{BC}$, $\overline{EF}$ are corresponding equal sides,

The triangle $ABC$ may be conceived to be moved [A 1] and placed so that the corner $B$ falls on $E$, the side $BC$ on $EF$, and since the angle $ABC$ is equal to the angle $DEF$, the side $BA$ upon the side $ED$.

Then since $\overline{BC}$ is equal to $\overline{EF}$, $C$ will fall upon $F$; and since $A$ is in $BA$, it will fall upon $ED$, or $ED$ produced.

Hence if a straight line through $F$ be twisted from the direction $FE$ towards the direction $FD$ [I. 4], measuring the angle $EFD$ [I. 5], and always intersecting $ED$, or $BA$, it will also measure the angle $BCA$. And since the angles $EFD$, $BCA$ are equal, it must reach $FD$ and $CA$ both at once.

Therefore $A$ must coincide with $D$, and the triangles be congruent.

Therefore *If two triangles......&c.* Q. E. D.

Corollary. *Hence if two angles of a triangle are equal, the two opposite sides are equal.*

For if the angle $ABC$ in the above proof had been equal to the angle $ACB$, the triangle $ABC$ might also have been moved so that $C$ fell upon $E$, and $B$ on $F$, and the triangles would also have been congruent so. Hence both the sides $\overline{AB}$, $\overline{AC}$ would be shown to be equal to $\overline{DE}$, and therefore to each other.

## Proposition XII.

*The greater side of every triangle is opposite to the greater angle,*

*And conversely, the side which is opposite to the greater angle is the greater side.*

For if one side $\overline{AC}$ of a triangle $ABC$ be greater than another $\overline{AB}$,

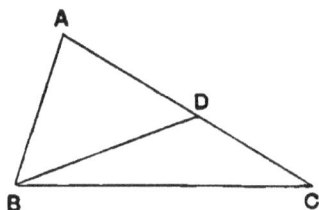

There must be some point $D$ between $A$ and $C$ such that $\overline{AD}$ is equal to $\overline{AB}$.   Join $BD$.                              [A 2.

Then because $D$ is in $AC$ between $A$ and $C$, the direction $BD$ is intermediate between and dependent upon the directions $BA$, $BC$.                                                    [I. 4.

Therefore the angle $ABC$ is greater than the angle $ABD$.
                                                       [I. 5 c.

But since the sides $\overline{AB}$, $\overline{AD}$ of the triangle $ABD$ are equal,
The angle $ABD$ is equal to the angle $ADB$.        [I. 10 c.

And as the angle $ADB$ is an exterior angle of the triangle $BCD$, it is greater than the interior opposite angle $BCD$.
                                                    [I. 8 c. (ii).

Much more therefore is the angle $ABC$ greater than the angle $BCD$, or $ACB$.

Therefore the greater side of every triangle is opposite the greater angle.

And conversely, the side which is opposite to the greater angle is the greater side.

For, if not, it must be either equal to or less than the other.

But if it were equal, the opposite angles would be equal [I. 10 c], whereas they are not.

And if it were less, the angle opposite to it would be the less (see above), whereas it is the greater.

Therefore *The greater side......&c.*                   Q. E. D.

## Proposition XIII.

*If two triangles have two sides of the one respectively equal to two sides of the other, but the angle between the directions of those sides greater in the one triangle than in the other, the third side of the former triangle shall be greater than the third side of the other.*

*And conversely, if the third side of the one triangle be greater than the third side of the other, the angle opposite to it in the former triangle shall be greater than the corresponding angle of the other.*

For if the sides $\overline{AB}$, $\overline{AC}$ of a triangle $ABC$ are respectively equal to the sides $\overline{DE}$, $\overline{DF}$ of a triangle $DEF$,

But the angle $BAC$ be greater than the angle $EDF$,

Then, of the sides $\overline{AB}$, $\overline{AC}$, one must be not less than the other. Suppose $\overline{AC}$ to be that one.

Through $A$ conceive a straight line to be twisted from $AB$ towards $AC$, always intersecting $BC$ in $X$ [I. 5], until it has described an angle equal to $EDF$.

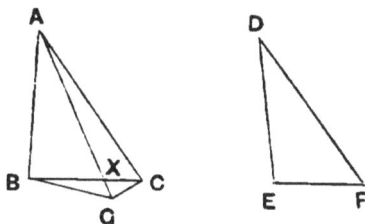

Then since the angle $EDF$ is less than the angle $BAC$, $X$ will be between $B$ and $C$. [I. 5 c.

Then since $\overline{AC}$ is not less than $AB$, the angle $ABC$ is not less than the angle $ACB$. [I. 12.

And since $AXC$ is the exterior angle of the triangle $ABX$,

The angle $AXC$ is greater than the interior opposite angle $ABC$. [I. 8 c. (ii).

Therefore it is greater than the angle $ACB$.

Therefore the side $\overline{AC}$ of the triangle $AXC$ is greater than the side $\overline{AX}$. [I. 12.

Hence if $G$ be the point in $AX$, in the same direction from $A$ as $X$ is, and at the same distance from it as $C$, the point $X$ will be between $A$ and $G$.

Therefore the direction $CX$ is intermediate between and dependent upon the directions $CA$ and $CG$. [I. 4.

D. 4

Therefore the angle $ACG$ is greater than the angle $XCG$.

[I. 5 c.

And similarly, because $X$ is between $B$ and $C$, the angle $BGC$ is greater than the angle $XGC$.                [I. 5 c.

But since $\overline{AG}$ is equal to $\overline{AC}$,

The angle $AGC$ is equal to the angle $ACG$.            [I. 10 c.

Therefore the angle $BGC$ is greater than the angle $ACG$ and much more therefore is it greater than $BCG$. (See above.)

Therefore the opposite side $\overline{BC}$ of the triangle $BGC$ is greater than the side $\overline{BG}$.                [I. 12.

But since the sides $\overline{AB}$, $\overline{AG}$ of the triangle $ABG$ are respectively equal to the sides $\overline{DE}$, $\overline{DF}$ of the triangle $DEF$, and the angle $BAG$ is equal to the angle $EDF$,

Therefore the triangles are congruent, and $\overline{BG}$ is equal to $\overline{EF}$.                [I. 10.

Therefore $\overline{BC}$ is greater than $\overline{EF}$.

And conversely, if $\overline{BC}$ is greater than $\overline{EF}$, the angle $BAC$ shall be greater than the angle $EDF$.

For, if not, it must be equal or less.

But if it were equal $\overline{BC}$ would be equal to $\overline{EF}$ [I. 10], which it is not,

And if it were less, $\overline{BC}$ would be less than $\overline{EF}$ (see above), whereas it is greater.

Therefore the angle $BAC$ must be greater than the angle $EDF$.

Therefore *If two triangles......&c.*                Q. E. D.


### PROPOSITION XIV.

*If two triangles have the three sides of the one respectively equal to the three sides of the other, they shall be congruent.*

For if one angle of the one were either greater or less than the corresponding angle of the other, the side opposite to this angle would be greater or less than the corresponding side of the other triangle, whereas it is equal to it.            [I. 13.

Therefore the angles between the directions of any two sides of the one triangle must be equal to the angle between those of the two sides equal to them of the other triangle.

Hence the triangles are congruent.                [I. 10.

Therefore *If two triangles......&c.*                Q. E. D.

### PROPOSITION XV.

*If two triangles have two sides of the one respectively equal to two sides of the other, and an angle of the one triangle opposite one of those sides equal to that opposite the equal side of the other triangle, then the angles of the two triangles opposite their other equal sides shall either be equal, or their sum shall be a straight angle.*

For if two sides $\overline{AB}$, $\overline{AC}$ of a triangle $ABC$ are respectively equal to two sides $\overline{DE}$, $\overline{DF}$ of a triangle $DEF$, and if the angles $ABC$, $DEF$ opposite the equal sides $\overline{AC}$, $\overline{DF}$ be equal,

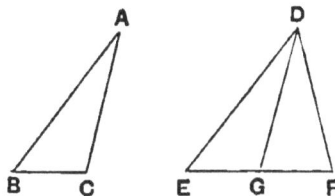

The triangle $ABC$ may be conceived to be moved [A 1] and placed so that the corner $B$ falls on $E$, $BA$ on $ED$, and $BC$ on $EF$, since the angle $ABC$ is equal to $DEF$.

And since $\overline{AB}$ is equal to $\overline{DE}$, $A$ will fall upon $D$.

And since $C$ falls on $EF$, either it falls upon $F$, and the triangles are congruent,

Or else it falls at some other point $G$ in $EF$, and $DGF$ is a triangle.

Of the two points $G$ and $F$, one of them, say $G$, must be between the other and $E$.

Then since the sides $\overline{DG}$, $\overline{DF}$ of the triangle $DGF$ are equal, the angle $DGF$ is equal to the angle $DFG$.      [I. 10 c.

But the sum of the angles $DGE$, $DGF$ is a straight angle.

[I. 6 c. (i).

Therefore the sum of the angles $DGE$, that is $ACB$, and $DFE$ is a straight angle.

Therefore *If two triangles......*&c.      Q. E. D.

Corollaries. *Hence* (i) *if besides the conditions in the enunciation it is known that the doubtful angles are either both greater, or both less, than a right angle, the triangles are congruent;* for then the sum of the doubtful angles cannot be a straight angle.

(ii) *If the known equal angles are either right angles, or greater than right angles, the triangles are congruent;* for since the sum of the known and the doubtful angle in either triangle is less than a straight angle [I. 8 c. (ii)], both the doubtful angles must be less than right angles.

(iii) *Or, if it is known that one of the doubtful angles is a right angle, the triangles are congruent;* for then in either case the other doubtful angle is a right angle and therefore the triangles are congruent.                                    [I. 11.

PROPOSITION XVI.

*Any two sides of a triangle are together greater than the third.*

For if *ABC* be any triangle,

Then if its sides are all equal, obviously any two of them together are greater than the third.

But if not, one of them, say $\overline{AB}$, must be not less than either of the others, and greater than one of them, say $\overline{AC}$.

Then there must be some point *D* between *A* and *B*, at the same distance from *A* as *C* is.   Join *CD*.            [A. 2.

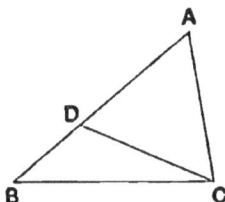

Then since the sides $\overline{AC}$, $\overline{AD}$ of the triangle *ACD* are equal,
The angles *ACD*, *ADC* are equal.                    [I. 10 c.

And therefore since their sum is less than a straight angle, each of them is less than a right angle.          [I. 8 c. (ii).

But the sum of the angles *CDB*, *CDA* is a straight angle.
                                                       [I. 6 c.

Therefore the angle *BDC* is greater than a right angle.

And the sum of the angles *BDC*, *BCD* is less than a straight angle.                                          [I. 8 c. (ii).

Therefore the angle *BCD* is less than a right angle.

Therefore the angle *BDC* is greater than the angle *BCD*.

Therefore the side $\overline{BC}$ of the triangle $BCD$ is greater than $\overline{BD}$. [I. 12.

To these unequals add the equals $\overline{AC}$ and $\overline{AD}$ respectively.

Therefore $\overline{BC}$, $\overline{CA}$ are together greater than $\overline{BD}$, $\overline{DA}$, that is, than $\overline{BA}$.

And since no side of the triangle is greater than $\overline{BA}$, any two sides of the triangle are together greater than the third.

Therefore *Any two......&c.* Q. E. D.

Corollary. *Hence a straight line is the shortest path between two positions.* For any other path may, within any assignable limits of accuracy, be supposed to be made up of a succession of short straight lines. And instead of traversing such a cornery path, it would, by this proposition, always be shorter to miss out the next coming corner, and to go straight to the next but one. *A fortiori* would it be shorter to miss out all the corners, and go straight from the one position to the other.

(NOTE. This is why the distance between two positions was defined as the amount of transference along a straight line between them.)

## PROPOSITION XVII.

*Any two of the angles between three independent directions are together greater than the third.*

For if $O$ be any position, and if straight lines $OA$, $OB$, $OC$ be conceived extending from it in any three independent directions,

Then, if the three angles between them are all equal, obviously any two of them are together greater than the third.

But if not, one of them, say $AOB$, must be not less than either of the others, and greater than one of them, say $AOC$.

Conceive a straight line [A 2] joining any two points $A$ and $B$ in the given directions from $O$, and conceive a straight line to revolve from $OA$ towards $OB$, always intersecting $AB$ in $D$, until the angle $AOD$ is equal to $AOC$. [I. 5.

Then $D$ will be between $A$ and $B$, since the angle $AOB$ is greater than $AOC$. [I. 5 c.

Let $C$ be the point in the third given direction from $O$, at the same distance from it as $D$ is. Join $CA$, $CB$. [A 2.

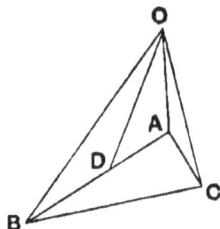

Then since the direction $OC$ is independent of the directions $OA$, $OB$; $OC$ cannot intersect $AB$, and $C$ is not in $AB$. [I. 4.

Therefore $ACB$ is a triangle. [D 12.

Therefore $\overline{AC}$, $\overline{CB}$ are together greater than $\overline{AB}$. [I. 16.

And since the sides $\overline{OA}$, $\overline{OC}$ of the triangle $OAC$ are respectively equal to the sides $\overline{OA}$, $\overline{OD}$ of the triangle $OAD$, and the angle $AOC$ is equal to the angle $AOD$,

The triangles are congruent, and therefore $\overline{AC}$ is equal to $\overline{AD}$. [I. 10.

But $\overline{AC}$, $\overline{CB}$ are together greater than $\overline{AD}$, $\overline{DB}$.

(See above.)

Therefore $\overline{CB}$ is greater than $\overline{DB}$.

But the sides $\overline{OC}$, $\overline{OB}$ of the triangle $OBC$, are respectively equal to the sides $\overline{OD}$, $\overline{OB}$ of the triangle $OBD$.

Therefore the angle $COB$ is greater than the angle $DOB$. [I. 13.

To each of these unequals add the equals $AOC$, $AOD$ respectively.

Therefore the angles $BOC$, $COA$ are together greater than $BOD$, $DOA$, that is, the angle $BOA$. [I. 5 c.

And since none of the three angles are greater than $BOA$, any two of them are together greater than the third.

Therefore *Any two......&c.* Q. E. D.

Corollary. *Hence the shortest way of twisting a straight line from one direction to another is as in measuring the angle between them.*

For any other way of twisting it may, within any assignable limits of accuracy, be supposed to be made up of a succession of short twistings as in measuring the angles between a series of directions. And instead of twisting it through such a succession of directions it would, by this proposition, always be shorter to miss out the next coming direction, and to twist it as if measuring the angle to the next but one. *A fortiori* would it be shorter to twist it at once as if measuring the angle from the first direction to the last.

(NOTE. This is why an angle was defined as the amount of twisting through a continuous series of directions intermediate between and dependent upon the extreme directions.)

PROPOSITION XVIII.

*The sum of the angles between any three independent directions is less than two straight angles.*

For if $O$ be any position and if straight lines be conceived through $O$ extending in any three independent directions $OA$, $OB$, $OC$ to points $A$, $B$, $C$,

Join $AB$, $BC$, $CA$;

Then since the direction $OA$ is independent of the directions $OB$, $OC$; $OA$ cannot intersect $BC$.       [I. 4.

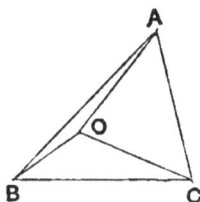

And since $AO$ does not intersect $BC$, the direction $AO$ is independent of the directions $AB$ and $AC$.       [I. 4.

Therefore the angles $OAB$, $OAC$ are together greater than $BAC$.       [I. 17.

Similarly the angles $OBA$, $OBC$ are together greater than $ABC$ and the angles $OCB$, $OCA$ are together greater than $BCA$.

Therefore the six angles $OAB$, $OAC$, $OBA$, $OBC$, $OCB$, $OCA$ together, are greater than the three angles of the triangle $ABC$,

That is, than a straight angle.       [I. 8.

To each of these unequals add the three angles $AOB$, $BOC$, $COA$,

Then the nine angles of the three triangles $OAB$, $OBC$, $OCA$, that is, three straight angles, are together greater than one straight angle together with the three angles $AOB$, $BOC$, $COA$.

Therefore these three angles are together less than two straight angles.

Therefore *The sum*......&c. Q. E. D.

## PROPOSITION XIX.

*One, and only one, straight line can be conceived to a given straight line from a given point without it, in a direction perpendicular to it.*

For if $O$ be any point outside a straight line $AB$,

Let $C, D$ be any two points in $AB$. Join $OC$.

Then if the angle $OCA$ is equal to the angle $OCB$, $OC$ is perpendicular to $AB$. [D 11.

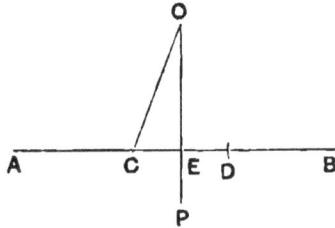

But if not, then since the sum of the angles $OCA$, $OCB$ is a straight angle [I. 6 c. (i)], one of them must be greater and the other less than a right angle. [D 9.

Let $OCB$ be the one that is less than a right angle.

Since every straight line through $O$ in a direction dependent on the directions $OC$ and $OD$ either intersects $AB$, or is parallel to it, [I. 4.

If a straight line $OP$ through $O$ revolve from $OC$ towards $B$, remaining in directions dependent upon $OC$ and $OD$, it will continue to intersect $AB$ on that side of $C$ on which $B$ is, until it is parallel to $AB$,

That is, until the sum of the angles $POC$, $OCB$ is a straight angle. [I. 9 c.

But since the angle $OCB$ is less than a right angle, before this $OP$ will reach a direction such that the sum of the angles $POC$, $OCB$ is a right angle. In this direction let it intersect $AB$ in $E$.

Then $OEB$ is the exterior angle of the triangle $OCE$ and

is therefore equal to the sum of the two interior opposite angles $EOC$, $OCE$ [I. 8 c. (i)], that is, to a right angle.

But the sum of the angles $OEB$, $OEA$ is a straight angle [I. 6 c.], that is, two right angles. [D 9.

Therefore the angles $OEB$, $OEA$ are equal, and $OE$ is perpendicular to $AB$. [D 11.

But no other straight line from $O$ to $AB$ can be so, for if one as $OD$ were, the angles $ODE$, $OED$ would be both right angles, and therefore their sum a straight angle, whereas being interior angles of a triangle their sum must be less than a straight angle. [I. 8 c. (ii).

Therefore *One, and only one,......&c* Q. E. D.

## Proposition XX.

*The shortest path to a straight line from a point without it is the straight line perpendicular to it.*

*And of all other straight lines to it from the given point, those which meet it at equal distances from the foot of the perpendicular are equal, and those which meet it nearer the foot of the perpendicular are shorter than those which meet it farther off.*

For if $O$ be any point outside a straight line $AB$,

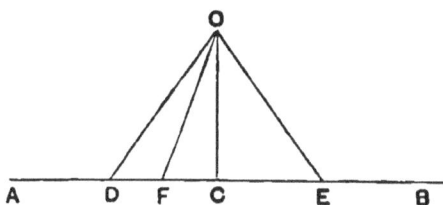

Conceive a straight line $OC$ from $O$ to $C$ in $AB$, in a direction perpendicular to $AB$. [I. 19.

Conceive a straight line from $O$ to any other point $D$ in $AB$.

Then since $ODC$ is a triangle, the sum of the angles $ODC$, $OCD$ is less than a straight angle. [I. 8 c. (ii).

But the angle $OCD$ is half a straight angle. [I. 6 & D 11.

Therefore the angle $ODC$ is less than half a straight angle, and therefore less than $OCD$.

Therefore the side $\overline{OC}$ of the triangle $OCD$ is less than the side $\overline{OD}$. [I. 12.

Therefore $\overline{OC}$ is the shortest straight line, and *a fortiori* the shortest path from $O$ to $AB$. [I. 16 c.

And if $E$ be any other point in $AB$ at the same distance from $C$ as $D$ is,

The sides $\overline{OC}$, $\overline{CD}$ of the triangle $OCD$ are respectively equal to the sides $\overline{OC}$, $\overline{CE}$ of the triangle $OCE$, and the angles $OCD$, $OCE$ are equal.

Therefore the triangles are congruent, and $\overline{OD}$ is equal to $\overline{OE}$. [I. 10.

And if $F$ be any point nearer to $C$ than $D$ or $E$ is, say on the same side as $D$,

Then it may be shown as above that the angle $OCF$ being a right angle, the angle $OFC$ is less than a right angle.

[I. 8 c. (ii).

Therefore the angle $OFD$ is greater than a right angle.

But the angle $ODF$ is less than a right angle, and therefore less than $OFD$.

Therefore the side $\overline{OF}$ of the triangle $OFD$ [I. 12] is less than the side $\overline{OD}$.

Therefore *The shortest path......&c.* Q. E. D.

## Proposition XXI.

*The straight lines which join those extremities of two equal and parallel straight lines which lie in opposite directions from the others, bisect one another;*

*And those which join those extremities which lie in the same directions from the others, are themselves equal and parallel.*

For if $\overline{AB}$, $\overline{CD}$ be two equal and parallel straight lines, the direction from $A$ to $B$ being the same as that from $C$ to $D$,

Then if a straight line through $A$ be conceived to revolve from the direction $AC$ towards $AD$ and so on, remaining always in directions dependent on $AC$ and $AD$, it will continue to intersect $CD$ on the same side of $C$ as $D$ is, until it is parallel to $CD$, that is, coincident with $AB$. [I. 4.

Therefore the direction $AD$ is intermediate between, as well as dependent upon, the directions $AC$, $AB$.

Therefore $AD$ must intersect $CB$ between $C$ and $B$, in $O$, say. [I. 4.

Similarly $CB$ intersects $AD$ between $A$ and $D$, in $O$.

Then, since the directions $AB$, $AD$ are respectively opposite to the directions $DC$, $DA$,

The angle $BAO$ is equal to the angle $CDO$.     [I. 7.

Similarly the angle $ABO$ is equal to the angle $DCO$,

And the sides $\overline{AB}$, $\overline{CD}$ of the triangles $OAB$, $ODC$ are equal.

Therefore the triangles are congruent.     [I. 11.

Therefore $\overline{AO}$ is equal to $\overline{OD}$ and $\overline{BO}$ to $\overline{OC}$, that is, $\overline{AD}$, $\overline{CB}$ bisect one another in $O$.

And since $\overline{OA}$ is equal to $\overline{OD}$ and $\overline{OB}$ to $\overline{OC}$, and the angles $AOC$, $DOB$ of the triangles $AOC$, $DOB$ are equal,

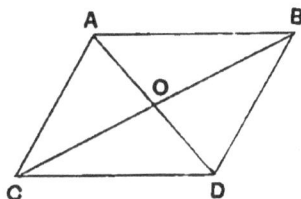

Therefore the triangles are congruent.     [I. 10.

Therefore $\overline{AC}$ is equal to $\overline{BD}$,

And the angle $ACB$ is equal to the angle $CBD$.

Hence the two straight lines $AC$, $BD$ each intersect each of the intersecting straight lines $AD$, $BC$ in separate points, and their directions $CA$, $BD$ on opposite sides of one of these straight lines, make equal angles with its opposite directions $CB$, $BC$.

Therefore they are parallel.     [I. 9.

And they have been shown to be equal.

Therefore *The straight lines......&c.*     Q. E. D.

## Proposition XXII.

*The perpendicular distance of any point in a straight line from a straight line parallel to it, is constant.*

For if $A$, $B$ be any two points in a straight line to which a straight line $CD$ is parallel,

Conceive the straight line $AC$ from $A$ to $CD$ perpendicular to $CD$,     [I. 19.

And let $D$ in $CD$ be in the same direction from $C$ as $B$ is from $A$, and at the same distance from it.

Then $\overline{AB}$, $\overline{CD}$ are equal and parallel straight lines.

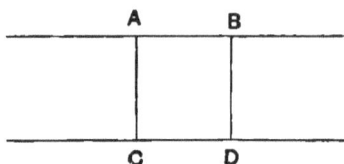

Therefore also $\overline{AC}$ and $\overline{BD}$ are equal and parallel.    [I. 21.

Therefore $\overline{BD}$ is the perpendicular to $CD$, and is equal to $\overline{AB}$.

Therefore *The perpendicular*......&c.        Q. E. D.

## PROPOSITION XXIII.

*The opposite sides and angles of a parallelogram are equal to one another.*

For if $ABDC$ be a parallelogram, $AB$ is parallel to $CD$.

[D 13.

Let $X$ in $CD$ be the point in the same direction from $C$ as $B$ is from $A$, and at the same distance from it,

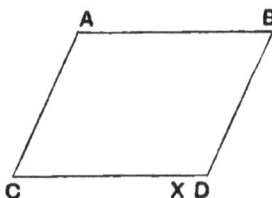

Then $\overline{AB}$, $\overline{CX}$ are equal and parallel straight lines.

Therefore $\overline{AC}$ and $\overline{BX}$ are equal and parallel straight lines.

[I. 21.

But $BD$ is parallel to $AC$.                  [D 13.

Therefore $BX$ and $BD$ extend in the same directions from $B$, and they are therefore one and the same straight line. [I. 1.

And as both $X$ and $D$ are in $CD$ also, they must be one and the same point.                  [I. 1.

Therefore $\overline{AB}$ and $\overline{CD}$ are equal, as also are $\overline{AC}$ and $\overline{BD}$.

And since $AB$ is parallel to $CD$, the sum of the adjacent angles of the parallelogram $BAC$, $ACD$ is a straight angle.

[I. 9 c.

And for the same reason the sum of the angles $ACD$, $CDB$ is also a straight angle.

Therefore the angle $BAC$ is equal to the angle $CDB$.

And similarly the angle $ACD$ is equal to the angle $DBA$.

Therefore *The opposite sides*......&c.                    Q. E. D.

Corollaries.  (i)  *Hence if it is possible for a point to move from one position to another by moving certain distances in two given directions successively, these movements may be taken in either order.*

(ii)  *Therefore also, if it is possible for a point to move from one position to another by moving certain distances in any number of given directions, these movements may be taken in any order whatever.*  For any desired change of order may be effected by repeatedly interchanging the order of two conse-cutive movements.  We see therefore that the dependence or independence of a given direction on certain other directions, is an intrinsic relation between the directions, and has nothing to do with the magnitude, or order of the transferences selected to test it by.

(iii)  *If two adjacent sides of a parallelogram are equal, all its sides are equal.*  Such a figure is called a rhombus.

(iv)  *If two adjacent angles of a parallelogram are equal, all its angles are equal, and they are all right angles.*  For the sum of two adjacent angles is a straight angle.  Such a figure is called a *rectangle*.  A figure which combines both the properties of a rhombus and a rectangle is called a *square*.

## BOOK II

## LEMMA.

*If each one of any group of directions is dependent upon a certain number of them, each of which is independent of the remainder in that number; then each one of the whole group of directions is dependent upon any similar number of independent directions chosen from among the group.*

Let the letters of the alphabet $ab......z$ denote the directions in the group, and let a certain number of them, $ab......k$ say, denote the certain number of directions, each of which is independent of the remainder, upon which all the directions are dependent.

The proposition will be proved by showing that for any one, $a$, of the independent directions $a—k$, we may substitute any one, $z$, of the remainder, and each one of the whole group will be dependent upon $b—k$ and $z$; provided only that $z$ is independent of $b—k$.

For by hypothesis, $z$ is dependent upon $a—k$.

Therefore it is possible to move a point from some position

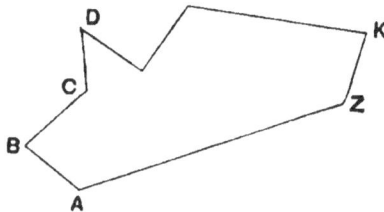

A to another Z, by a straight line $AZ$ extending in the direction

$z$, or by a succession of straight lines $AB$, $BC$, $CD$......$KZ$ in the successive directions $a$, $b$, $c$......$k$. [D 5.

The distances to be moved along some of these lines may perhaps be zero, but since the direction $z$ is independent of the directions $b$—$k$, the distance along $AB$ cannot vanish, for else it would be possible to move a point from $A$ to $Z$ by straight lines in directions $b$—$k$ only.

Therefore it is possible to move a point from $A$ to $B$ by a straight line in the direction $a$, or else by straight lines $AZ$, $ZK$......$DC$, $CB$ in directions $z$ and $k$......$c$, $b$, respectively.

Therefore the direction $a$ is dependent upon the directions $z$ and $b$—$k$. And if $x$ be any other direction whatever in the group, then by hypothesis a movement in direction $x$ may be replaced by movements along straight lines in directions $a$—$k$. But we have already seen that a movement along a straight line in direction $a$ may be replaced by movements along straight lines in the directions $z$ and $b$—$k$.

Therefore a movement in direction $x$ may be replaced by movements along straight lines in directions $z$ and $b$—$k$, for the order of the movements is indifferent. [I. 23 c. (ii).

Therefore *If each one*......&c. Q. E. D.

NOTE. We see from this proposition that directions may be classified in groups, as follows:

(*a*) Groups dependent upon one independent direction.

Such a group contains only two directions, the given direction, and the one opposite to it. But there are any number of such groups (see A 3).

(*b*) Groups dependent upon two independent directions.

Such a group contains all the directions in which a straight line extends while revolving to measure a straight angle. There are also any number of such groups (see A 3).

(*c*) Groups dependent upon three independent directions.

Material space, as we ordinarily conceive it, consists of only one such group, and it is with such a space that we are here concerned (see A 3).

Such groups may, for brevity, be spoken of as groups of one, two, or three independent directions.

### Additional Axiom.

III. Space may be conceived as extending from every position in it in three and only three independent directions.

[Since (Ax. II.) a straight line may be conceived to extend in any given direction from any given position, all directions conceived by virtue of Axiom III. belong to *one group* of three independent directions, and the directions in which space extends are the same from every position in it.]

### Further definitions.

16. A continuous series of positions extending from each one of them in a complete group of one, two, or three independent directions, is called a '*spread*.' If the directions in which it extends from every position in it are the same, it is called a *regular* spread.

17. A spread of one independent direction is called a *line*, and a regular spread of one independent direction a *straight* line (see also **D 4**). If however the directions in which it extends vary gradually from point to point in the line, it is called a *curved* line.

18. A spread of two independent directions is called a *surface*, a regular spread of two independent directions a *plane* surface, or merely a *plane*. If however the directions, or some of them, in which it extends vary gradually from point to point in the surface, it is called a *curved* surface.

19. The spread of three independent directions called *space* is a regular spread (see **A 2, 3**). Any limited portion of it is called a geometrical *solid*.

20. A *dimension* is a measurement made upon a figure with a view to determining its size.

21. *Linear* dimensions are measures of the extension of portions of spreads of one independent direction, that is, of lines.

22. Among linear dimensions, *length, breadth*, and *thickness*, are measured along three straight lines in independent directions. *Circumference* is measured round a closed line surrounding the figure, &c.

23. *Superficial* dimensions are measures of the extension of portions of spreads of two independent directions, that is, of surfaces.

24. Among superficial dimensions *plane areas* are measured over planes, and *curved areas* over curved surfaces, &c.......

25. *Volumetric* dimensions or *volumes* are measures of the extension of portions of spreads of three independent directions, that is, of solids. Since by Axiom III. we have only conceived one such spread, we need only conceive one kind of volume.

26. If two planes intersect in a straight line, the angle between directions in each perpendicular to that straight line is called the *angle between the planes*.

27. A *tetrahedron* is a figure consisting of six straight lines joining four points, which are not in one plane, two and two.

28. A *parallelepiped* is a figure consisting of twelve straight lines forming the intersections of three pairs of parallel planes.

(The terms tetrahedron and parallelepiped are often also used to denote the envelopes formed by the planes in which these straight lines are, or the geometrical solids which they enclose.)

29. Two positions are said to be *on opposite sides* of a point when they lie in opposite directions from it; in other words, when the point is in the straight line joining the positions, and between them;

So two positions are said to be on opposite sides of a line or surface if the straight line joining them intersects the line or surface once between them;

But if the line or surface intersect the straight line joining the positions twice between them, each of the positions is on the opposite side of the line or surface to a position in the straight line between the two intersections, and they are therefore on the same side of the line or surface as each other.

So generally, two positions are said to be on the same, or opposite sides of a line or surface, according as the straight line joining them intersects that line or surface an even or an odd number of times, between them.

### NOTE.

With reference to the definitions of the words parallel and perpendicular, the reader is requested to observe that they apply equally to

planes, or to a plane and straight line. Thus one plane is parallel to
another if it extends in all the same directions as it. So a plane is parallel
to a straight line if it extends in its two directions; but it would not be
accurate on this definition to speak of the straight line as parallel to a
plane. From the definition of perpendicular, it is not strictly accurate
to talk of anything but a direction as perpendicular to anything. But
since if two straight lines are at right angles, each of the directions of
either is perpendicular to the other, the straight lines may be called
perpendicular. So with a straight line and a plane. But it would be
wrong to speak of two planes which are at right angles, as perpendicular to
each other.

A plane is denoted by three letters denoting three points in it which
are not in a straight line (see II. 3).

## PROPOSITION I.

*A plane may be conceived through any given position extend-
ing in any two given independent directions.*

For if $O$ be a given position,

Then straight lines may be conceived through $O$ to points
$A$, $B$, in the given directions from $O$.                    [A 2.

And a transverse straight line may be conceived joining
$AB$.                                                          [A 2.

And straight lines may be conceived from $O$ to every point
in $AB$, and also one through $O$ parallel to $AB$.            [A 2.

These straight lines are all in directions dependent upon
the directions $OA$, $OB$, and their directions include all which
are so dependent.                                             [I. 4.

And the directions form a continuous series. For clearly
those of the straight lines which intersect $AB$ form a continuous

series, since the positions in $AB$ form a continuous series. [D 4.
And the remaining directions, namely, the two in which $AB$ extends, form part of the continuous series, the straight lines in which intersect any other transverse straight line $AB'$ not parallel to $AB$.

Therefore a continuous series of positions has been conceived extending from $O$ in a complete group of two independent directions, dependent on the directions $OA$, $OB$.

Moreover, it extends from any other point $P$ in it in the same, and no other directions.

For, since the straight line $OP$ is wholly in the series of positions, it extends from $P$ in the directions $PO$ and $OP$.

And if it extends from $O$ in any other direction to any point $X$,

Through $P$ conceive a straight line $PY$ parallel to $OX$, and let $Y$ be any point in it. Join $OY$. [A 2.

Then since it is possible to move a point from $O$ to $Y$ by the straight line $OY$, or else by the straight lines $OP$, $PY$,

The direction $OY$ is dependent on the directions $OP$, $PY$,
[D 5.

That is, on the directions $OP$, $OX$.

But each of these is dependent on the directions $OA$, $OB$.

Therefore the direction $OY$ is dependent on the directions $OA$ and $OB$. [Lemma.

Therefore the point $Y$ is in the series of positions.

Similarly any point in $PY$ is in the series of positions. Therefore the series extends from $P$ in the direction $PY$, that is, the direction $OX$, and so in any direction in which the series extends from $O$.

And it extends in no other direction.

For if it extends in any direction $PZ$, and $Z$ be any point in this direction from $P$, in the series of positions,

Then the direction $OZ$ is dependent on the directions $OA$, $OB$, as also is the direction $OP$.

Therefore the direction $PZ$, which is dependent on the directions $OP$, $OZ$, is dependent on the directions $OA$, $OB$.
[D 5, Lemma.

Therefore *A plane may be conceived* &c. Q. E. D.

Corollaries. (i) *Hence an unterminated straight line may be conceived to revolve in a plane about a fixed point in itself, describing an angle, and before it has completed a straight angle it will have passed once, and only once, through every point in the plane.*

(ii) *Hence also any number of straight lines through a fixed point, in a plane, may be conceived to revolve similarly,* and if they revolve at the same rate their situations relative to each other will remain unaltered.

(iii) *And therefore any portion of a plane may be conceived to revolve in that plane, round any fixed point in it, and may therefore be moved to any part of the plane without any alteration of shape or size, so far as this can be accomplished by translation and such rotation alone.*

## PROPOSITION II.

*If a straight line intersect a plane which is parallel to it, or if a straight line intersect any plane in two separate points, it is wholly in the plane.*

For if a straight line intersect a plane which is parallel to it, a point moving from the intersection in either of the directions in which the straight line extends would move along the straight line, and also along the plane, for both extend in these two directions. [D 10.

Therefore the straight line is wholly in the plane.

And if a straight line intersect a plane in two separate points, both straight line and plane extend in the directions from one of the points to the other, and the straight line extends in no other directions.

Therefore the plane is parallel to the straight line and therefore, as above, it wholly contains it.

Therefore *If a straight line......&c.* Q. E. D.

## PROPOSITION III.

*One plane, and only one, can be conceived through—*
   (i) *any three points which are not in a straight line;*
   (ii) *any straight line, and a point without it;*
   (iii) *any two straight lines which intersect one another;*
   (iv) *any two parallel straight lines.*

For (i) since the three points are not in a straight line,

The directions from one of them to the other two are independent directions [D 5], and therefore a plane may be conceived through this point extending in those independent directions, and it will contain the other two points.     [II. 1.

And any plane containing the three points must also extend from the first point in the directions of the other two, and therefore be one and the same plane.

(ii) Any two points being taken in the straight line, one and only one plane can pass through these and the given point, and as the straight line intersects this plane in two separate points it is wholly in it.     [II. 2.

(iii) Besides their point of intersection a point may be selected in each of the intersecting straight lines, and one and only one plane conceived through these three points (see above) which will contain both the straight lines.     [II. 2.

(iv) Any point being selected in one of the parallel straight lines, one and only one plane may be conceived through this point and the other straight line (see above), and since this plane is parallel to the first straight line, and the straight line intersects it, it is wholly in it.     [II. 2.

Therefore *One plane, and only one......&c.*     Q. E. D.

## PROPOSITION IV.

*Two unterminated straight lines in a plane either intersect or are parallel.*

For if *AB, CD* be any two unterminated straight lines in a plane, which are not parallel,

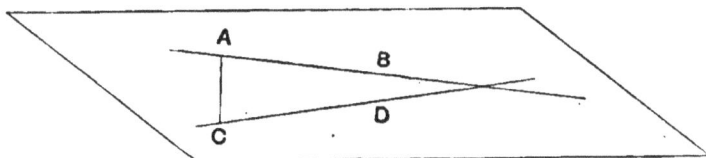

Conceive a straight line [A 2] joining any point *A* in one and any point *C* in the other.

Then the plane extends from *A* only in the independent directions *AB, CD*, and in directions dependent upon them.     [D 18.

Therefore since it extends in the direction $AC$, this direction is dependent upon the directions $AB$, $CD$.

Therefore since it is possible to conceive a point to move from $A$ to $C$ along $AC$ it is possible to conceive it to do so along straight lines in the directions $AB$, $CD$. [D 5.

Therefore the straight lines $AB$ and $CD$, which are in those directions, must intersect each other. ·

Therefore *Two unterminated......&c.* . Q. E. D.


PROPOSITION V.

*Any straight line, to which a plane is not parallel, intersects that plane in a single point.*

For if $AB$ be any straight line, and $CDE$ a plane which is not parallel to it,

Conceive a straight line joining any point $A$ in $AB$ to any point $C$ in $CDE$.

Then since the plane is not parallel to $AB$, the direction

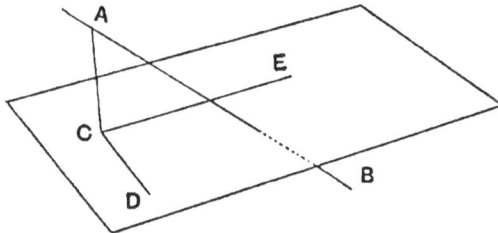

$AB$ is independent of the directions $CD$, $CE$ in which the plane extends.

Therefore the direction $AC$ is not independent of the three independent directions $AB$, $CD$, $CE$. [A 3.

Therefore, since it is possible to conceive a point to move from $A$ to $C$ along $AC$, it is possible to conceive it to do so along straight lines in the directions $AB$, $CD$, $CE$. [D 5.

The first of these motions is along the straight line $AB$, and the others along straight lines in the plane $CDE$.

Therefore $AB$ and $CDE$ intersect,

And since the plane $CDE$ is not parallel to $AB$, $AB$ is not wholly in it. [D 18.

Therefore it cannot intersect it in more than one point. [II. 2.

Therefore *Any straight line......&c.* Q. E. D.

## PROPOSITION VI.

*Two parallel planes cannot intersect one another,*

*And any two planes which are not parallel intersect in a straight line.*

For if two planes which are parallel had a common point, each would extend from that point in all the same directions, and they would therefore be one and the same plane. [D 10.

But if they are not parallel, let $A$ be a point in one of them, $ABC$, which is not in the other, $DEF$.

Then the plane $DEF$ cannot extend in both the independent

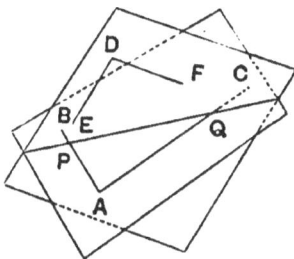

directions $AB$, $AC$, for it is not parallel to the plane $ABC$. [D 10.

Let it therefore not extend in the direction $AB$.

Therefore a straight line $AB$ will intersect it in a single point $P$. [II. 5.

And if the plane $DEF$ extends in the direction $AC$, a straight line through $P$ in this direction is wholly in both planes. [II. 2.

But if not, a straight line $AC$ also intersects the plane $CDE$ in a single point $Q$ [II. 5], which is not the same as $P$, since $A$ is the only point common to $AB$ and $AC$ [I. 1], and $A$ is not in the plane $CDE$.

Therefore a straight line $PQ$ intersects both planes in two separate points, and is therefore in both of them. [II. 2.

And no point outside this straight line can be in both planes, or they would be one and the same plane. [II. 3.

Therefore *Two parallel planes......*&c. Q. E. D.

Corollary. *Two planes cannot enclose a space.*

## PROPOSITION VII.

*If a plane intersect two parallel planes it does so in two parallel straight lines.*

For if it intersects them at all, it does so in two straight lines. [II. 6.

And these straight lines being in one plane must either intersect or be parallel. [II. 4.

But they cannot intersect, since they are in parallel planes, which do not intersect. [II. 6.

Therefore they are parallel.

Therefore *If a plane......&c.*　　　　　　Q. E. D.

## PROPOSITION VIII.

*If a direction be perpendicular to each of two intersecting straight lines, it is perpendicular to a plane determined by them.*

For if a direction be perpendicular to each of two straight lines *AO, BO* which intersect in *O,*

Through *O* conceive a straight line *COC′* in this direction and let *C, C′,* be any two points in it equidistant from *O.*

And if *OP* be any other direction in which the plane extends which is determined by *OA, OB,*

Through *P* a straight line may be conceived in the plane

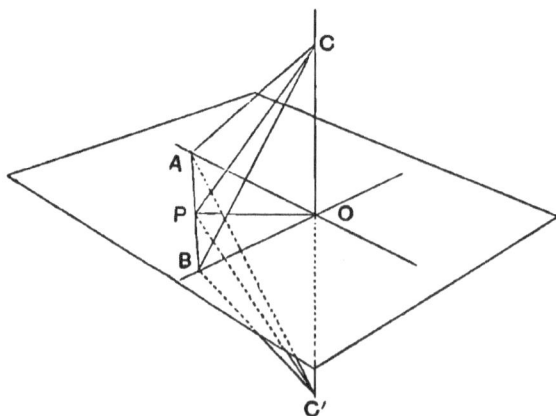

which is not parallel to *OP, OA,* or *OB* [II. 1 *c.* (i)], and which therefore does not pass through $O$, but which does intersect $OA$, and $OB$, in two separate points $A$ and $B$. [II. 4.

Join the points $C$, $C'$ to the points $A$, $B$, and $P$.

Then since the direction $OC$ is perpendicular to $AO$, the direction $AO$ is also perpendicular to $COC'$.

Therefore $A$ is equidistant from the points $C$, $C'$, which are equidistant from $O$, the foot of the perpendicular. . [I. 20.

. Similarly it may be shown that $B$ is equidistant from $C$, $C'$.

Therefore in the triangles $ACB'$, $AC'B$ the sides $\overline{AC}$, $\overline{CB}$ are respectively equal to the sides $\overline{AC'}$, $\overline{C'B}$ and the side $\overline{AB}$ is common.

Therefore the triangles are congruent. [I. 14.

But if the triangle $ACB$ were placed so as to coincide with $AC'B$, $\overline{PC}$ would coincide with $\overline{PC'}$.

Therefore $\overline{PC}$ is equal to $\overline{PC'}$.

And in the triangles $PCO$, $PC'O$ the sides $\overline{CO}$, $\overline{C'O}$ are also equal, and the side $\overline{PO}$ is common.

Therefore the triangles are congruent [I. 14] and the angle $POC$ is equal to the angle $POC'$, and therefore each of them is a right angle. [I. 6. c. (i).

Therefore the inclination of $OC$ to $OP$ is the same as to $OA$ or $OB$. [I. 6. c. (ii).

Thus the direction $OC$ is equally inclined to every direction in which the plane determined by $OA$, $OB$ extends.

Therefore the direction $CO$ is perpendicular to the plane. [D 11.

Therefore *If a direction......&c.* Q. E. D.

## Proposition IX.

*In a plane, through any given point in it, one and only one straight line can be conceived in a direction perpendicular to a given straight line in the plane.*

If the given point is without the given straight line the proposition has already been proved [I. 19]; for the perpendicular from the point to the straight line is in the plane containing them.

But if not, a straight line through the given point may be conceived to revolve in the given plane, starting from the directions of the given straight line, and describing an angle. And before completing a straight angle it will have passed through every point in the plane, and therefore have extended in every direction in which it extends.                                   [II. 1. c. (i).

And during this revolution, when it has twisted through a right angle, its directions will be equally inclined to the opposite directions of the given straight line [I. 6. c. (ii)]; that is, they will be perpendicular to it once [D 11], and once only, during the revolution.

Therefore *In a plane*......&c.                          Q. E. D.

## PROPOSITION X.

*Through any given point one, and only one straight line may be conceived in a direction perpendicular to a given plane.*

For if *O* be the given point and *OA* a straight line through *O* in any direction in which the given plane extends,

Through *O* conceive a plane [II. 1] *OAB* parallel to the given plane, and in it conceive the straight line *OB* in a direction perpendicular to *OA*.                          [II. 9.

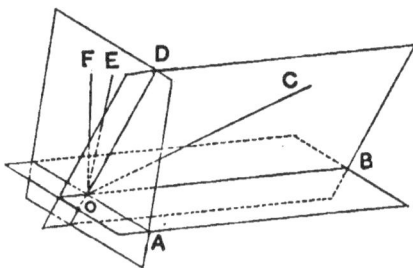

Conceive any point *C* in a direction from *O* independent of the directions *OA, OB*.                          [A 3.

Through *OC, OB* conceive a plane, and in this plane conceive the straight line *OD* perpendicular to *OB*.          [II. 3.

Then since the planes *OBA, OBC,* are not parallel, they intersect only in the straight line *OB*.                          [II. 6.

Therefore the straight lines *OD, OA* are in independent directions.

Through *OA, OD* therefore conceive a plane [II. 3], and in this plane conceive the straight line *OE*, in a direction perpendicular to *OA*.                          [II. 9.

Then since the direction *OB* is perpendicular to the straight lines *OA* and *OD*, it is perpendicular to the plane *AOD*, and therefore to the straight line *OE* in this plane. [II. 8.

Therefore *OE* is perpendicular to *OB*, and it is so also to *OA*, and therefore to the plane *AOB*, and to the given plane to which *AOB* is parallel. [II. 8.

And if any other straight line *OF* through *O* could be in a direction perpendicular to the given plane, or to *AOB*,

A plane could be conceived through *OE* and *OF* [II. 3], which would not be parallel to the plane *AOB*, and would therefore intersect it in a straight line [II. 6], to which the directions *OE*, *OP* would be perpendicular. [II. 8.

But the straight lines *OE*, *OF* are in one plane with this intersection, and therefore cannot both be in directions perpendicular to it. [II. 9.

Therefore *Through a given point......&c.* Q. E. D.

## PROPOSITION XI.

*Through any given point one, and only one plane may be conceived to which a given direction is perpendicular.*

For if *O* be the given point and *OA* a straight line through *O* in the given direction,

Through *O* conceive two more straight lines, [A 2], such

that the directions *OA*, *OB*, *OC* are independent. [A 3.

Through *OA*, *OB*, and *OA*, *OC* conceive two planes [II. 1], and in them conceive the straight lines *OD*, *OE* in directions perpendicular to *OA*. [II. 9.

Since the planes $AOB$, $AOC$ are not parallel, they only intersect in the straight line $OA$. [II. 6.

Therefore the directions $OD$, $OE$ are independent. Conceive a plane through $OD$, $OE$. [II. 3.

Then $OA$ is perpendicular to both $OD$ and $OE$, and therefore to the plane $DOE$. [II. 8.

And if it were perpendicular to any other plane through $O$, and $F$ were any point in this plane and not in $DOE$,

Conceive a plane through $OA$, $OF$ [II. 3]. Then this plane would not be parallel to the plane $DOE$, and therefore it would intersect it in a straight line $OG$. [II. 6.

And $OA$ would be perpendicular to $OF$ and $OG$.

But $OF$ and $OG$ are in the same plane as $OA$, and therefore could not both be perpendicular to it. [II. 9.

Therefore *Through any given point*......&c. Q. E. D.

## PROPOSITION XII.

*The shortest path to a plane from a point without it is a straight line in a direction perpendicular to the plane.*

*And of all other straight lines to it from the given point, those which meet it at equal distances from the foot of the perpendicular are equal, and those which meet it nearer the foot of the perpendicular are shorter than those which meet it farther off.*

For if $O$ be any point outside a plane $ABC$,

Conceive a straight line $OP$ [II. 11] in a direction perpendicular to $ABC$. Then since $ABC$ is not parallel to $OP$ it must intersect it in a single point $P$. [II. 5.

If $OA$ be any other straight line to $A$ in the plane, from $O$,

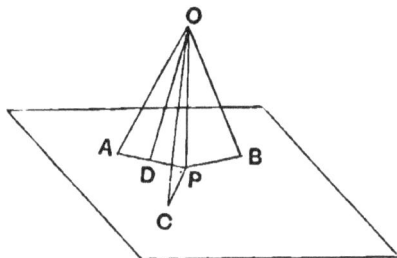

then $AP$ is in the plane [II. 2], and therefore $OP$ is perpendicular to $PA$. [D 11·

Therefore $\overline{OP}$ is less than $\overline{OA}$. [I. 20.

Thus $\overline{OP}$ is the shortest straight line, and *a fortiori* the shortest path from $O$ to the plane.

And if $B$ be any other point in the plane, at the same distance from $P$ as $A$ is,

Then the sides $\overline{AP}$, $\overline{PO}$ of the triangle $APO$ are equal to the sides $\overline{BP}$, $\overline{PO}$ of the triangle $BPO$ respectively, and the angle $APO$ is equal to the angle $BPO$, since $OP$ is perpendicular to the plane $APB$.

Therefore the triangles are congruent, and $\overline{OA}$ is equal to $\overline{OB}$. [I. 10.

And if $C$ be any point in the plane nearer to $P$ than $A$,

Some point $D$ in $AP$ is at the same distance from $P$ as $C$ is, and is therefore nearer than $A$ is.

Therefore the straight line $\overline{OD}$ is equal to $\overline{OC}$ (see above), and less than $\overline{OA}$. [I. 20.

Therefore also $\overline{OC}$ is less than $\overline{OA}$.

Therefore *The shortest path*......&c. Q. E. D.

PROPOSITION XIII.

*The perpendicular distance of any point in a straight line or plane from a plane which is parallel to it, is constant.*

For if $A$, $B$ be any two points in a straight line or plane to which a plane $CDE$ is parallel,

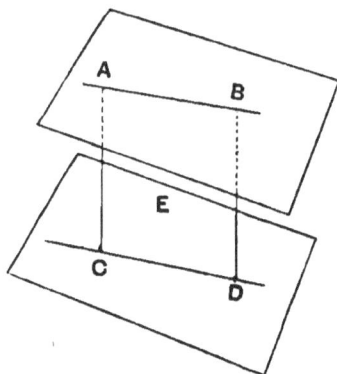

Conceive a straight line $AC$ in a direction perpendicular to the plane $CDE$. [II. 11.

Then as $CDE$ is not parallel to $AC$, it will intersect it in a single point $C$. [II. 5.

Through $C$ conceive a straight line $\overline{CD}$ equal and parallel to $\overline{AB}$. [A 2.

Then since the plane $CDE$ is parallel to $AB$, $D$ will be in this plane.

Join $BD$. Then $\overline{BD}$ is also equal and parallel to $\overline{AC}$.

[I. 21.

Therefore it is the perpendicular from $B$ to the plane $CDE$.

[II. 10.

And it is equal to $\overline{AC}$, the perpendicular from $A$.

Therefore *The perpendicular distance......&c.* Q. E. D.

## PROPOSITION XIV.

*An unterminated plane may be conceived to revolve round any fixed straight line in it, describing an angle; and before completing a straight angle it will have passed once and only once through every point in space.*

For if $AB$ be any fixed straight line in a plane and $O$ any point in it,

Conceive a plane $COD$ through $O$, to which $AB$ is perpendicular. [II. 11.

Then a straight line $XOY$ through $O$ may be conceived to revolve in this plane, describing an angle. [II. 1 $\overset{\bullet}{c}$.

And in every position of it a plane may be conceived through it and $AB$. [II. 3.

And since as $XOY$ revolves in the plane $COD$, $XOY$ is

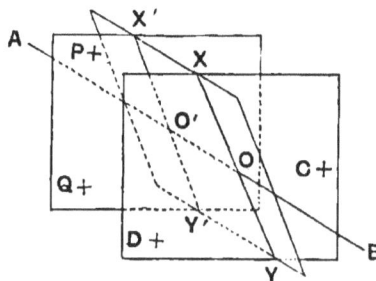

always in a direction perpendicular to $AB$, the intersection of any two situations of the plane.

Therefore the plane $XOA$ in revolving describes an angle,

[D 26], namely the same as that described by $XOY$ in the plane $COD$.

And if $P$ be any point in space,

Through $P$ conceive a plane $PO'Q$ parallel to the plane $COD$. [II. 1.

As this is not parallel to $AB$, it will intersect it in a single point $O'$. [II. 5.

And the plane $XOA$ will intersect the parallel planes $COD$, $PO'Q$ in parallel straight lines $XOY$, $X'O'Y'$. [II. 7.

Hence as the plane $XOA$ revolves describing an angle, $X'O'Y''$ will revolve in the plane $PO'Q$ describing the same angle.

But as $X'O'Y'$ revolves, before completing a straight angle it will pass once and only once through every point in the plane, and therefore through $P$. [II. 1 c. (i).

Hence as the plane $XOA$ revolves, before completing a straight angle it will pass once through $P$.

And it cannot pass through $P$ except when $X'O'Y'$ does, for if it did, it would coincide with the plane $PO'Q$ [II. 3] which it does not, since it contains $AB$, which is not wholly in that plane, but perpendicular to it.

Therefore *An unterminated plane......&c.* Q. E. D.

Corollaries. (i) *Hence also any number of planes through a fixed straight line may be conceived to revolve similarly*, and if they revolve at the same rate their situations relative to one another will remain unaltered.

(ii) *And consequently any geometrical solid may be conceived to revolve in space round any fixed straight line in space, and may therefore be moved from any one part of space to any other, without alteration of shape or size, so far as this can be accomplished by translation and such rotation alone.*

(iii) *Hence any plane figure may be conceived to be moved into any given plane, so that any point in it shall fall upon a given point in that plane, any straight line through that point, on a given straight line through the given point, and any other point in it on either side of the given straight line in the given plane.* For any point in the figure may be moved to the given point by a translation, that is, without altering the direction from any one point to any other in the figure. And then the plane of the figure and the given plane will intersect in a straight line,

[II. 6], by a rotation round which the plane figure may be brought into the given plane. The figure may now be rotated in the given plane [II. 1 c. (iii)] round the given point till any straight line through it coincides with the given straight line. And lastly, if any other point in the figure is on the wrong side of the given straight line, it may be brought on to the right side by revolving the plane figure through a straight angle round the given straight line.

This corollary shows that none of the movements of figures we have conceived in the past propositions have been inconsistent with axioms II. and III.

## Proposition XV.

*It is impossible to pass from one side of an unterminated surface in space to the other, without passing through the surface.*

For if $O$, $P$, be any two points on opposite sides of an unterminated surface in space,

Then a straight line $OP$ intersects the surface an odd number of times between $O$ and $P$.                     [D 29.

The proposition will be proved by showing that however $P$ may move in space, if it does not pass through the surface, the number of intersections between $O$ and $P$ will always remain odd, and can therefore never vanish.

For in whatever direction $P$ may move, unless in the direction of $O$ or the opposite direction, $OP$ will commence to generate a plane.                                          [II. 1.

Therefore however $P$ moves, except in the direction of $O$ or the opposite, $OP$ will describe a spread which from every point in it extends in the same directions as a plane, that is, a surface.

And if $P$ only moves in the direction of $O$ and the opposite, the number of intersections between $O$ and $P$ will remain unaltered, unless $P$ passes through one of them, that is, through the given surface.

Therefore as $P$ moves $OP$ will generate a surface bounded by the path of $P$, the point $O$, and two situations of the straight line $OP$.

And from any of the intersections of $OP$ with the given surface, this surface and that described by $OP$ extend in all the directions in which two planes would ;

Their intersection therefore extends in all the directions in which the intersection of two planes would;

That is, in two opposite directions.                    [II. 6.

Therefore the intersection of the two surfaces consists of one or more lines which are not terminated, unless by the boundaries of the surface generated by $OP$.

Hence if $A$ be one of the intersections between $O$ and $P$, as $OP$ revolves $A$ will describe a continuous line on the surface generated by $OP$, and the intersection $A$ can only vanish if at some point $B$ this line bends back as $OP$ moves forward,

And in this case, before reaching $B$, $OP$ must have inter-

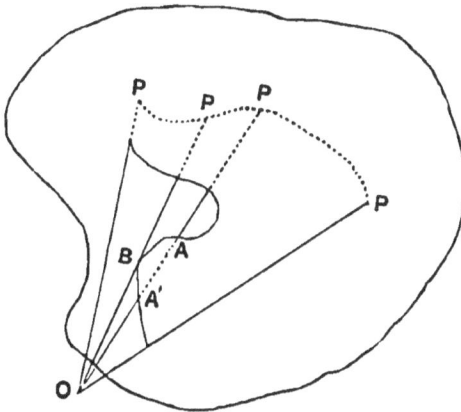

sected the continuous line traced by $A$ in a second point $A'$, which therefore vanishes together with $A$ as $OP$ passes the point $B$.

And the line described by $A$ cannot be terminated in the path described by $P$, unless $P$ passes through the given surface, nor can it be terminated at $O$, since $O$ is not in the given surface.

Thus as $OP$ revolves the intersections can only appear or disappear two at a time.

Therefore *It is impossible......&c.*                    Q. E. D.

Corollary.   *In the same way it may be shown that in a plane it is impossible to pass from one side to the other of an unterminated line without passing through the line.*

NOTE.   In this proposition contacts of the second order must be counted double intersections, of the third order triple, &c.   Also if the given surface intersects or touches itself it must be counted double at such points, &c. It is not necessary to discuss these special cases here, but they might perhaps be a reason for omitting this proposition from an elementary text-book.

## Proposition XVI.

*If two tetrahedra have their six corresponding sides respectively equal, they shall be equal in every respect, that is to say, the angles between the directions of corresponding sides shall be equal, and the angles between corresponding plane faces shall also be equal.*

*And if their corners are similarly situated with respect to each other in space, they shall be congruent to each other.*

For since the corresponding sides of the tetrahedra are respectively equal, they consist of four pairs of corresponding triangles, whose three sides are respectively equal.

Therefore these corresponding triangles are congruent to each other. [I. 14.

Therefore also the angles between the directions of corresponding sides which meet at a corner are equal, for they are angles of corresponding triangles.

Conceive one of the tetrahedra to be moved and placed so that one of its triangles coincides with the corresponding triangle of the other, which is congruent with it. [A. 1.

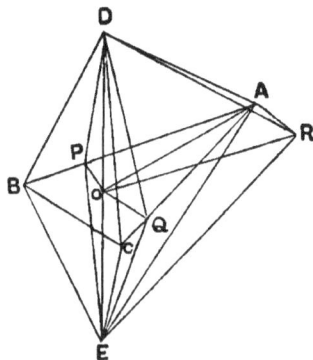

Let $ABC$ be this triangle, and $D$, $E$ the remaining corners of the two tetrahedra.

Now since the triangles $ADB$, $AEB$ are congruent (see above),

If they were applied to each other so as to coincide, one

perpendicular [I. 19] could be conceived from their common corner $D$, or $E$, to the side $AB$, meeting it in one point $P$.

Therefore even though $D$, $E$ do not coincide, the perpendiculars from them on $AB$ meet it in the same point $P$.

Similarly, conceive perpendiculars from $D$ and $E$ on $AC$ meeting it in one point $Q$.

And in the plane $ABC$ conceive straight lines through $P$ and $Q$ in directions perpendicular to $AB$, $AC$ respectively.

[II. 9.

These straight lines are not parallel, for if they were, in the plane $ABC$ two straight lines $AB$, $AC$ through $A$ would be in directions perpendicular to the same straight line, which is impossible.

Therefore they intersect in some point $O$. [II. 5.

Join $DO$, $EO$,

Then as the direction $AB$ is perpendicular to both $PD$ and $PO$ it is perpendicular to a plane containing them [II. 8], and therefore to $DO$ in this plane.

Similarly the direction $AC$ is perpendicular to $DO$.

Therefore the direction $DO$ is perpendicular to both $AB$ and $AC$, and therefore to the plane $ABC$. [II. 8.

Similarly $EO$ is perpendicular to the plane $ABC$.

But only one straight line may be conceived through a given point, in a direction perpendicular to a given plane. [II. 10.

Therefore $DOE$ is one straight line.

And since the triangles $DAB$, $EAB$ are congruent,

(see above)

The perpendiculars $\overline{DP}$ and $\overline{EP}$ are equal.

And the side $\overline{PO}$ is common to the two triangles $DPO$, $EPO$.

Also the angles $DOP$, $EOP$ opposite the equal sides $\overline{DP}$, $\overline{EP}$ are equal, and moreover they are right angles.

Therefore the triangles are congruent [I. 15. $c$ (ii)], and therefore the angles $DPO$, $EPO$ are equal.

But these angles are the angles between the planes $DBA$, $EBA$, since $PD$, $PO$, $PE$ are perpendicular to $AB$. [D 26.

So it may be shown that the angles between any two corresponding planes are equal.

Through $A$ conceive a straight line $AR$ parallel to $BC$, to any point $R$. Join $DR$, $ER$, $OR$.

Then since $DOE$ is perpendicular to the plane $ABC$ (see above), and $OR$ is in this plane, it is perpendicular to $OR$.

Therefore $RO$ is the perpendicular from $R$ on $DOE$.

But it was shown above that the triangles $DPO$, $EPO$ are congruent. Hence $D$ and $E$ are equidistant from $O$.

Therefore $\overline{RD}$ is equal to $\overline{RE}$.                          [I. 20.

And also $\overline{AD}$ is equal to $\overline{AE}$, and $\overline{AR}$ is common to the triangles $DAR$, $EAR$.

Hence the triangles are congruent [I. 14], and therefore the angle $DAR$ is equal to the angle $EAR$.

Therefore the angles between the directions of corresponding sides of the tetrahedra which do not meet at a corner, as the sides $BC$, $AD$ and $BC$, $AE$, are also equal.

And since it has been shown above that $\overline{DO}$ is equal to $\overline{EO}$, and is in the same straight line with it,

If the corners of the tetrahedra are similarly situated with respect to each other, $D$ and $E$ will be on the same side of $O$, and will therefore be one and the same point.

Therefore the tetrahedra are congruent.

Therefore *If two tetrahedra......&c.*                          Q. E. D.

Corollaries.  (i) *Hence if two tetrahedra have three sides meeting at one corner of the one respectively equal to the corresponding three of the other, and have the three corresponding angles between their directions respectively equal, they shall be equal in every respect.*  For the third sides of each of the three triangles meeting at the corner, will be respectively equal also.                          [I. 10.

(ii) *Or if they have the six angles of each, between the corresponding sides which meet at two corners, respectively equal, and the sides between those corners equal, they shall be equal in every respect.*  For the other four sides meeting at those two corners are respectively equal [I. 11], and therefore the tetrahedra are equal in every respect.                          [c. (i) above.

(iii) *Hence a direction may be determined with reference to two given independent directions by giving its respective inclinations to them, if it is also known which side it is of a plane determined by those given directions.*

## PROPOSITION XVII.

*The four diagonals of a parallelepiped, that is, the four straight lines joining its opposite corners, intersect in a single point, which bisects each of them.*

For a parallelepiped consists of the intersections of three pairs of parallel planes. [D 28.

And each pair intersects each of the other planes in a pair of parallel straight lines. [II. 7.

Hence the parallelepiped consists of twelve straight lines, four of which extend in each of three independent directions,

Or of twelve straight lines joining eight points, two and two, three meeting at each point and extending from it in three independent directions.

Hence if $AB$, $CD$, $EF$, $GH$ be four parallel sides of a

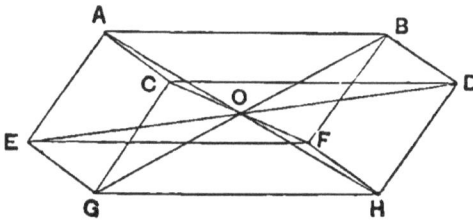

parallelepiped, $AE$, $BF$, $CG$, $DH$ a second four, and $AC$, $BD$, $EG$, $FH$ the remainder,

$ABDC$ is a parallelogram. [D 13.

Therefore $\overline{AB}$ is equal to $\overline{CD}$. [I. 23.

Similarly $\overline{CD}$ is equal to $\overline{GH}$.

Therefore $\overline{AB}$, $\overline{GH}$ are two equal and parallel straight lines.

Therefore $\overline{AH}$ and $\overline{BG}$ bisect one another, in $O$ say. [I. 21.

Similarly $\overline{AC}$, $\overline{FH}$ are equal and parallel straight lines.

Therefore $\overline{AH}$ and $\overline{CF}$ bisect one another, and they therefore do so in $O$ also.

Similarly $\overline{ED}$ passes through $O$ and is bisected by it.

Therefore *The four diagonals......&c.* Q. E. D.

TABLE OF THEOREMS WITH THEIR EQUIVALENTS IN EUCLID.

| Proposition | Euclid's equivalent | Proposition | Euclid's equivalent |
|:---:|:---:|:---:|:---:|
| I. 1 | I. D 35 and A 10 | I. 21 | I. 33 |
| 2 | none | 22 | none |
| 3 | none | 23 | I. 34 |
| 4 | none | II. 1 | none |
| 5 | none | 2 | I. D 7 |
| 6 | I. 14 and c. A 11 | 3 | XI. 2 (?) |
| 7 | I. 15 | 4 | I. D 35 |
| 8 | I. 32 and c. I. 16, 17 | 5 | none |
| 9 | I. 29 and c. I. 27, 28 | 6 | XI. D 8, P. 3 |
| 10 | I. 4 and c. I. 5 | 7 | XI. 16 |
| 11 | I. 26 and c. I. 6 | 8 | XI. 4 |
| 12 | I. 18, 19 | 9 | I. 11 |
| 13 | I. 24, 25 | 10 | XI. 6, 11, 12 |
| 14 | I. 8 | 11 | XI. 5 |
| 15 | (analogous to VI. 7) | 12 | none |
| 16 | I. 20 | 13 | none |
| 17 | XI. 20 | 14 | none |
| 18 | XI. 21 | . 15 | none |
| 19 | I. 19 | 16 | none |
| 20 | none | 17 | none |

Proofs are given of all Euclid's theorems (and such of his problems as are of theoretical importance,) in his first and eleventh books, with the exception of those in his first book which deal with plane areas, and of one in his eleventh book which treats of proportion, and of one other theorem; this one is his I. 21, a theorem of no importance. It may however be easily deduced, after my I. 16, if an equivalent condition is substituted for the implied one that the point 'inside the triangle' is in the same plane as the triangle, or as it stands after my II. 3.

# PART III.

## ON THE APPLICABILITY OF THE FOREGOING SUBJECTIVE GEOMETRY TO THE GEOMETRY OF MATERIAL SPACE.

## CHAPTER I.

HAVING thus established the foundations of subjective geometry of three independent directions, we may now confront the question: Is there a corresponding objective geometry; and if so, does it follow the same laws? The question whether there is or is not an objective world at all, is a metaphysical one of the abstrusest character, and involves the discussion of the logical basis of the inductive method in its most abstract form. We need hardly, however, attack this question directly here, as no practical man will doubt that the outer world has a real objective existence. But it may be useful to consider briefly how we attained to our knowledge of its objectivity.

I think it will be generally agreed that the lower forms of animal, and possibly the higher forms of vegetable life, though they can hardly have formed the conception of an objective external world, do feel certain subjective impressions of pleasure and pain, of heat and cold, and so on. At what point in the scale of existence automatic and reflex actions are divided from voluntary ones, and whether even voluntary actions necessarily imply a consciousness of objective existence, is open to dispute. But those, at any rate, who believe in Evolution can scarcely doubt that, after a certain period of development with only a subjective consciousness, after an epoch of many generations of life with experiences only recognised as subjective, the perception that these were

due to an objective environment must have sprung into being.
And as such a perception must have been of great advantage
to the percipient, it would only require natural selection to
confirm and extend it. Thus we might have, and probably
have, an instinctive consciousness of the objectivity of our
environment, without this consciousness constituting a direct
apprehension of an objective fact; and this consciousness therefore does not add anything to our reasons for believing that
our environment is objective. Such a consciousness might
indeed seem to Kant to constitute an 'apodictic' truth, but it
does not constitute a necessary one.

Granted then that there is an objective universe, is its
geometry identical with the subjective geometry we have been
considering ? The question we have to answer is merely this :
Are the axioms of that subjective geometry true in the material
universe ?

The first axiom, that any geometrical figure may be conceived to be moved from any one part of space to any other,
except in so far as it is restricted by the other axioms, without
its shape or size being altered, is subjectively a truism; but
considered objectively it raises several difficult points. We
may indeed take it to be sufficiently established by induction
that a material body may be moved to any part of space ; and
further we have doubtless sufficient ground for believing that
any forces required to produce such motions are accounted for
otherwise than by any change of size or shape which the body
is required to undergo by the nature of space, on being moved.
But we have no *a priori* reason for assuming that such a change
of size or shape would require force to produce it, and to discuss
whether it is produced or not we are thrown back upon first
principles.

Size and shape depend on distances and inclinations respectively, which in our subjective geometry were measured by
amounts of transference and twisting. In virtue of the first
axiom we were enabled to substitute direct comparison by the
method of superposition for these methods of measurement.
But if the axiom is not objectively true, presumably the size
and shape of the figure applied would alter, on being moved,
in the same way as that to which it was applied, and so the
method of superposition would not prove that two congruent

figures were of the same size or shape when in different parts of space. If we attempt to define 'amount of transference' or of 'twisting' strictly we can only do so by the aid of Newton's first law of motion, which would introduce the conception of time. For just as I have already shown that the words 'in a straight line' in his law can only be construed as meaning what I mean by the phrase 'in a constant direction,' so the words 'uniform motion' mean nothing but 'motion with a constant amount of transference in any given portion of time.' And the only measure we have of time is the motion of a body not acted on by any forces (the rotation of the earth is the ordinary standard, and very closely approximates to an ideal one). Thus we should seem to be defining in a vicious circle; and though this is not quite the case, for the law of motion implies a *consistency* between the measures of distance or inclination, and time, deduced from it, yet the criterion it affords is difficult to apply as a practical test. However, since from this law of motion the most remote results have been deduced, and have been found to correspond with objective facts, its truth, at least in so far as it asserts that a body not acted on by any effective forces moves a constant distance in a constant time, may fairly be assumed. And so the first axiom of subjective geometry may be translated into the first objective fact of the geometry of material space.

It will be convenient to consider the third axiom next, and to discuss whether material space is a spread of three independent directions, that is, whether it does extend from every position in it in a complete group of three independent directions. That it does do so is established by a pure induction. The objectivity of my arm being granted, I know I could extend it in three independent directions, and in any direction dependent upon them, from my present position in space, or from any position I have ever occupied (where I was not obstructed by material objects). And by induction I infer that the rest of space is like those parts of it with which I am acquainted. Other instances and facts might be adduced proving the same thing; but the logical proof must be of the same nature, namely, a generalisation from a limited number of known instances, to an unlimited number of unknown ones, and though of course it is amply good enough for all practical purposes, it establishes only an objective fact, not a necessary truth. It applies with equal force to prove that material space extends from no position in it in less than three independent directions, or from no position in more. Thus we have established the second objective fact on which to base the geometry of material space.

I have already pointed out in Part I that the remaining axiom may be paraphrased into a form of which the objective counterpart would be : Material space extends from every position in it in *the same* directions. Now we have already granted that from every position in it it extends in three independent directions, and in all directions dependent upon them. Therefore if from any position $B$ in it it extended in any direction in which it does not extend from a position $A$, this direction could not be dependent upon the three independent directions in which material space extends from $A$. And we should therefore have four independent directions to deal with.

Now I am aware that there exists, not only among the ordinary public, but even more among geometricians, a rooted prejudice against what they would call ʻGeometry of Four Dimensions,ʼ but which I prefer to call geometry of four independent directions. If therefore any of my readers is of opinion that ʻA Fourth Dimensionʼ or a Fourth independent direction is *a priori* inconceivable, and an impossibility ;—then I have no more to say. He has *ipso facto* granted the objective truth of my remaining axiom, and consequently admitted that my subjective geometry may also be applied objectively to the geometry of material space—which is all I desire to prove. The remainder of this book is not for him : if he reads it, and thinks it transcendental folly, let him at least not presume to criticise what, by his own confession, he does not understand.

# CHAPTER II.

I HAVE already had occasion to point out that, if we are careful to reason with formal accuracy, it is not essential that we should be able to picture to ourselves clearly every link in the deductive chain, and that intelligible results may be truly deduced by the aid of symbols whose denotation is unintelligible to us. It is therefore no argument against the theory of geometry of four independent directions which I am about to advance, to say that a fourth independent direction is inconceivable. Personally, I believe it to be by no means inconceivable, and I shall subsequently give my reasons for this belief. But lest the reader should imagine that the validity of my investigation into the foundations of the science of geometry in any way depends upon this conceivability, I put the investigation, or the essential part of it, first, and the explanation afterwards.

If such an investigation has never been made before, it is simply because the foundations of geometry, especially those of what is generally spoken of as 'geometry of three dimensions' have never before been laid down with formal accuracy. We have already seen that even in his first book Euclid assumes several things which are not stated in his postulates or axioms. But in his eleventh book he seems to abandon all attempts at laying down his premises with formal accuracy, and simply appeals to the supposed intuitive knowledge of his reader. The first three propositions of his eleventh book are most extraordinary examples of such slipshod reasoning, and what is more extraordinary still, is that scarcely any effort has been made by more modern geometricians to improve them! The existence, even, of the first proposition is only due to a strange confusion of ideas. Up to this Euclid has been drawing straight lines freely without ever considering whether they might or might not have loose ends hanging out of the plane in which he tacitly assumes all his former figures to have been drawn. It is only now dawning on him that the existence of this plane, which is really nothing but the pedagogue's black-board, is a matter which he ought not so airily to have taken for granted. Again, the second proposition contains simply *no* formal reasoning whatever, but is merely an appeal to the experience or intuitive knowledge of the reader. Take the second paragraph[1] verbatim, thus—" Let any plane pass through the straight line *EB*......." Where is the postulate by which it is allowed to conceive, or draw, any plane through a given straight line? "......and let the plane be turned round *EB*......." How do we know that a plane can be turned round a fixed straight line? "......produced if necessary......" (i.e. the plane, not the straight line). But there is no postulate about producing planes "...... until it pass through the point *C.*" Now in this last respect alone, as far as I know, has it occurred to any one materially to improve this proposition. In a recent work[2] an ' axiom ' is specially interpolated to cover this difficulty. The ' axiom ' in question I prove in a theorem (II. 14), namely, that before a plane revolves through a straight angle round a fixed straight line in it, it will pass through every point in space. Again, in the next proposition, what right has Euclid to assume, as he does, that if two planes cut one another their common section must be a line, at all? Why should it not be a single point? in which case his proof would break down. It is the adroit manner in which Euclid begs this question which saves him from having to lay down anything corresponding to my third axiom.

The geometry of three dimensions, as usually put forward, is not then a formal deductive science, though apparently it satisfies the wants of most men. But what I maintain about my geometry of three independent directions is that it is strictly formal,—the truth of every proposition in it depends solely on the truth of the stated premises. And if one of those premises were altered we could see at once which of the propositions would have to be altered, and we should probably be able to determine the alteration required. If the third axiom

---

[1] Todhunter's *Euclid*, Macmillan, 1875.
[2] *Geometry in Space*, by Mr R. C. J. Nixon.

is to be altered, it is at once evident that no change is produced in the first book, as the axiom was not even stated till the beginning of the second. And it is easy to trace which of the propositions in the second book depend upon it.

They are namely these :—V., that part of VI. which asserts that *any* two planes which are not parallel intersect in a straight line, VII., for without the axiom the intersections might only be single points, X. and XI., when they assert that *only* one plane or straight line may be conceived, XIV., when it asserts that the revolving plane passes through *every* point in space, XV. ; and from XVI. we may remove the restriction that the corners of the tetrahedra must be similarly situated in space, in order that they may be congruent.

Now let us substitute for the third axiom the following postulate,—

Let it be granted that the positions of points may vary in four, but not in more than four independent directions.

And let us see what happens to the above seven propositions. Thus it appears that we are no longer able to assert in proposition V. that the direction $AC$ cannot be independent of the three independent directions $AB$, $CD$, $CE$. But if $AB$ intersects the plane $CDE$, it will be so dependent. Therefore $AB$ does not necessarily intersect the plane. It may pass from one side to the other without intersecting it at all. That is to say, the positions of points which appear to us the same, even though picturing to ourselves a three dimension model, instead of a diagram on paper, may in reality be different ; just as when looking at an ordinary diagram representing a three dimension figure, positions appear the same in it which may yet be different in reality. Thus within the bounding surface of a solid of three dimensions we must allow for an infinite number of positions which do not form part of that solid, but lie away from it in a fourth independent direction. Therefore the whole of the 'three dimension space' we are in the habit of considering is only an infinitesimal portion of the space of four independent directions assumed by the postulate, just as a plane is only an infinitesimal portion of a space of three independent directions.

Let us therefore invent a special name for such limited · portions of the new space, and lay down the following new definition.

A spread of three independent directions is called a *form*, a regular spread of three independent directions a *regular form*.

The spread of four independent directions, the new space, is by the second axiom a regular spread, for this axiom is, of course, still to hold good.

Thus the space we have hitherto been considering is a regular form, and to make the seven propositions enumerated above still hold true we have only to prefix the words, " In a regular form " to their enunciations, just as the enunciations of several other propositions in the same book commence with the words, " In a plane."

It will not be necessary to give all the elementary propositions of geometry of four independent directions *in extenso* as I have done those of three. For the logical processes by which they are established are in most cases obvious extensions of those used in Book II., and the reader can if necessary readily enough supply them himself. But in the subjoined syllabus of enunciations one or two of the most important proofs have been sketched out.

## SYLLABUS OF ENUNCIATIONS FOR BOOK III.

[In this syllabus the letters r. f. mean regular form, p. plane, s. l. straight line, i. d. independent direction.  P. postulate].

I.   *One, and only one, r. f. can be conceived through any given point, extending in any three given i. d's.*

This is only another way of enunciating II. 14.

II.   (i) *If a p. intersect a r. f. in three points not in a s. l.; or*

(ii)   *if a s. l. intersect it in two separate points; or*

(iii)   *if either a p. or a s. l. to which it is parallel intersect it at all, then they shall be wholly in it.*

III.   *One and only one r. f. can be conceived through*

(i)   *four points not in one p.,*

(ii)   *a p. and a point without it,*

(iii)   *two p's. which intersect in a s. l., and so on.*

IV.   *Any s. l. to which a r. f. is not parallel intersects it in a single point.*

For if $A$ be any point in a s. l. $AB$ to which a r. f. $CDEF$ is not parallel, the four directions $AB$, $CD$, $CE$, $CF$ are independent. Therefore the direction $AC$ is not independent of them (P). That is, it is possible to move from $AC$ by a s. l. $AC$, or by s. l's. in directions $AB$, $CD$, $CE$, and

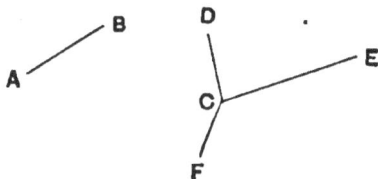

$CF$ (D 5). The first of these motions is in the s. l. $AB$, the others in the r. f. $CDEF$. Therefore they intersect. And they cannot intersect in more than one point. [III. 2.

V. *Any p. to which a r. f. is not parallel intersects it in a single s. l.*

For if $A$ be a point in the p. $ABC$ not in the r. f. $DEFG$; since the r. f. is not parallel to the p. it is not parallel to both $AB$, $AC$. Therefore one of them, $AB$, intersects it in a single point $P$ (III. 4). If the r. f. is parallel to $AC$, a s. l. through $P$ parallel to $AC$ is in both p. and r. f.

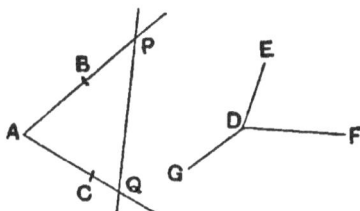

(III. 4). If not $AC$ intersects the r. f. in a single point $Q$, and the s. l. $PQ$ is in both p. and r. f. And they cannot intersect in any point outside $PQ$ (III. 2).

VI. *Two parallel r. f's. cannot intersect; and any two which are not parallel intersect in a p.*

For if $ABCD$, $EFGH$ be two r. f's. which are not parallel, and $A$ be any point in the first which is not in the second, then $EFGH$ is not

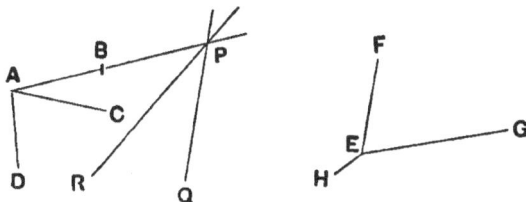

parallel to all of $AB$, $AC$, $AD$. Therefore it intersects at least one of them in a single point, $P$ (III. 4). And *EFGH* is not parallel to either of the p's. *ABC*, *ABD*. Therefore it intersects each of them in a s. l. *PQ*, *PR* (III. 5). Therefore also the p. *PQR* is in both r. f's. And they cannot intersect in any point outside this p. (III. 2).

**VII.** *Any two p's. extending in four i. d's. intersect in a single point.*

For if *ABC*, *DEF* be two p's. which extend in four i. d's., and if *A* be any point in the first which is not in the second, then the direction *AD* is not independent of the four i. d's. *AB*, *AC*, *DE*, *DF* (P). Therefore it is possible to move from *A* to *D* by a s. l. *AD* or by s. l's. in these four directions (D 5). But the first two of these motions are in *ABC*, and the other two in *DEF*. Therefore the planes intersect, and they cannot intersect in more than one point, else a r. f. might be conceived through them [III. 3], in which case they would not extend in four i. d's.

Corollary. *If two p's. extend in only three i. d's. they do not intersect at all, unless both are in one r. f.*

For if they only extend in three i. d's. either they are both in one r. f. or two parallel r. f's. may be conceived through them. And as these parallel r. f's. do not intersect, neither do the p's. they contain.

**VIII.** *If each of two i. d's. in one plane be at right-angles to each of two in another, any direction in the one p. is perpendicular to the other p.*

This follows at once from the definition of perpendicular. Two such planes may be called perpendicular to each other.

**IX.** *One, and only one, p. may be conceived through a given point perpendicular to a given p.; and it intersects in a single point.*

For if *D* be the given point and *ABC* the given p. conceive a s. l. *DP* to a point *P* in a s. l. *AB* in the p. *ABC*, perpendicular to *AB* (I. 19); and in the p. *ABC* conceive *PC* in a direction perpendicular to *AB* (II. 9),

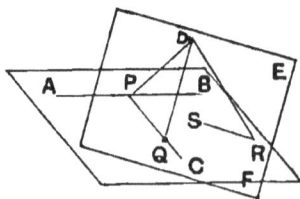

and conceive *DQ* to a point *Q* in *PC*, in a direction perpendicular to *PC* (I. 19). Then *DQ* is perpendicular to the p. *ABC* (II. 8). And if *R* be any point in a direction from *D* independent of *QD*, *AB* and *AC* in the same way *RS* may be conceived to *S* in *ABC*, in a direction perpendicular to it. Then since *R* is not in a r. f. *ABCD*, *SR* is in a direction independent of *QD*, *AB*, *AC*. Therefore it is not parallel to *QD*. Therefore a p. *DEF* may be conceived through *D* in the directions *QD*, *SR* which is perpendicular to the p. *ABC* (III. 8 and 7), and which intersects it in a single point *Q*. And if any other p. through *D* were perpendicular to *ABC*, it must also contain *DQ*, and therefore be in one r. f. with the p. *DEF* (III. 5). But the p. *ABC* would intersect this r. f. in a single s. l. which therefore would be perpendicular to both the p's. through *DQ*, and in one r. f. with them, which is impossible (II. 11).

X.   *If a direction be perpendicular to each of three i. d's. in a r. f. it is perpendicular to the r. f.*

A mere extension of my II. 8, or Euclid's XI. 4.

XI.   *One, and only one, s. l. may be conceived through a given point in a direction perpendicular to a given r. f. And conversely one and only one r. f. may be conceived through a given point to which a given direction is perpendicular.*

XII.   *The perpendicular s. l. is the shortest path from a given point to a given r. f. And, of all other s. l's....and* so on.

XIII.   *An unterminated r. f. may be conceived to revolve about any fixed p. in it describing an angle. And before it has completed a straight angle, it will have passed once, and only once, through every position in the new space of four i. d's. we are considering.*

For if *ABC* be the fixed p. and *O* any point in it, through *O* one, and only one, p. *DEF* can be conceived perpendicular to *ABC* (III. 9) and a s. l. *XOY* in this p. may revolve describing an angle, and a r. f. may always be conceived through this and *ABC* (III. 3), which will revolve with it, describing the same angle. And a p. may be conceived through any point *P* in the new space, parallel to the p. in which *XOY* revolves (III. 9), which will intersect the p. *ABC* in a single point *O'* and the r. f. *XOAB* intersects this p. in a s. l. parallel to *XOY*. And so on, as in II. 14.

Corollaries.   (i) *Hence any number of r. f's. may be conceived to revolve similarly about the same p.*

(ii) *And consequently any figure of four dimensions may be conceived to revolve about a fixed p. in it.*

Hence two tetrahedra which have their six sides respectively equal must be congruent. For their bases being made to coincide, the vertex of one may be revolved about its base till it coincides with the vertex of the other.   In doing so the moving vertex will however have to leave the r. f. containing the other tetrahedron, only returning to it every time it has revolved through a straight angle.

XIV.   *If two pentangular bodies have their ten sides respectively equal, they shall be equal in every respect, and if their corners are similarly situated with respect to each other, they shall be congruent.*

By a pentangular body I mean the figure formed by the straight lines joining five points not in a r. f. two and two.   Thus there are $\dfrac{5 \cdot 4}{1 \cdot 2}$, that is 10 sides.   The proof is precisely analogous to II. 16.

# CHAPTER III.

In the last chapter the elements of geometry of four independent directions were carried as far as I had previously carried those of geometry of three; and I think it must now be evident to the reader that they might without difficulty be carried further without our having any clear idea of what the words 'a fourth independent direction,' mean. I shall therefore now attempt to show how an idea can be formed of the meaning of those words, and if even I fail, the reader will recognise that my main argument is in no way damaged by the failure.

Perhaps the most striking change introduced into geometry by conceiving variations of position in a fourth independent direction, is the possibility of getting past an unterminated surface without passing through it. To take the simplest case, conceive a material plane, and a material point moving towards it, and getting past it. What does one mean by asserting that at some moment the point must have intersected the surface? That at that moment the position of the point must have been identical with some position now occupied by a point belonging to the plane. The plane and point being material bodies this would seem self-evident, but so far from being a necessary truth, it is only one of the experimental facts upon which we found our belief in the second Objective Fact of material geometry; and if the plane and point are not supposed to be material, but merely mental concepts, the assertion no longer has a leg to stand on. We are undoubtedly able to conceive a space of three independent directions, and if we imagine points not to be material, we may conceive any number of them apparently occupying the same position in such a space, but their positions in reality differing in a direction in which we have not power to move material points. In just the same

7—2

way if we were to look at a motionless view through a hole in a screen, it would have just the appearance of a perfect picture, painted on a plane surface. We should see its length and breadth but be unable to judge its depth. We might see two lines cross each other, without being able to assert positively that they intersected, for at their apparent point of intersection one might be at a greater depth than the other. To get a true idea of what was before us we should need to regard the view, or picture, from another point of view—under ordinary circumstances indeed, we do regard every picture from two points of view, namely from our two eyes. So also if we actually had before us a figure of four dimensions, we could not find it out, because we could not move so as to get another view of it. But as the four dimension figures we have to deal with are not actually before us, but are merely mental concepts, we may conceive ourselves, mentally, to walk round them, or to twist them round so as to expose various views to us, each of which views we can mentally picture to ourselves as a model in three dimensions. As these models are merely geometrical projections, their shape can be determined by the methods of formal geometry we have indicated above. For we have seen that the material space we commonly conceive is, geometrically, a regular form. The differences of position in a figure of four dimensions, which we do not perceive in a model of only three, are dimensions of the figure measured in a fourth independent direction, which is perpendicular to this regular form. Hence the model is found by projecting every point in the real figure by a straight line perpendicular to material space. Thus the apparent length of any straight line in the three dimension model is its true length, multiplied by the cosine of the angle it makes with its projection in material space; and so on, just as in ordinary orthogonal projection.

I am rather shy of here introducing a device, which however I have found most useful myself, to aid me in conceiving figures of four dimensions. For I have found it very difficult to prevent people from taking the shadow for the substance, from thinking that what I mean to put forward merely as an illustration, is in some way intended as physical fact. But having warned the reader, I must run the risk, and trust to him not to misunderstand me.

Let us then picture to ourselves figures of four dimensions
by conceiving models, or diagrams in material space (which for
this purpose we conceive as a regular form), which diagrams
are orthogonal projections of the true figures, and let us
conceive the distance of every point in the figure from a
regular form parallel to material space (but all on one side of
the figure), to be represented in the diagram by the degree of
blackness of the corresponding point in the diagram. The
fourth dimension of the figure, over and above the length
breadth and thickness of the diagram, we may call its solidity;
for just as its thickness is the difference in height of its top
and bottom, so its new dimension is the differences of the new
measurements of its opposite boundaries in the new direction,
that is the difference between two degrees of blackness, which
we may accurately represent by a certain degree of density
given to each point. Care must be taken however to distinguish between 'solidity' which is a dimension of a body, and
the 'blackness' which we have used to represent a measurement
from a fixed regular form—which is in fact a coordinate—and
which we may call 'concentration.' Thus if we consider a three
dimension solid, such as a sphere, it is clear that corresponding
to any vertical line through it, it has only one thickness, but
its surface has two heights. So with a body of four dimensions
it will, at any point in its three dimension diagram, have only
one solidity, but its boundary will have two concentrations.
A representation of the body by its solidity cannot therefore
give a perfect idea of its shape, to do which a double diagram,
of concentrations of its two boundaries, is required. But any
such a diagram of concentrations may be regarded as a diagram
of solidities of a body bounded by one irregular boundary,
and by the regular form from which the concentrations are
measured.

Thus in a diagram of concentrations a line will be a line
of varying blackness, unless material space is parallel to it.
A straight line will not only necessarily appear straight in the
diagram, but its degree of blackness will vary uniformly from
one end to the other. Two parallel straight lines will appear
as parallel straight lines, of shade varying at the same rate and
in the same direction. A plane in the same way will appear as
a plane shaded in parallel lines of shade whose blackness varies

uniformly in one direction. Similarly a regular form will appear as a cloud in space, stratified in parallel plane strata whose blackness varies uniformly in one direction. Hence, even if two lines appear to intersect, they do not really do so, unless at their apparent intersection they are equally shaded. Thus a straight line appearing to intersect a plane need not necessarily do so. Two planes may appear to intersect, but, unless there are points of equal shade in their apparent intersection, they do not do so in reality. As however their shadings vary uniformly, they will vary uniformly along the straight line in material space along which they appear to intersect. Hence, unless the lines of shade are both parallel to this line, in which case there is no variation of shade along it, or unless the shadings in both planes vary at the same rate in the same direction along it, one of the shadings will somewhere along the apparent intersection overtake the other. And at this single point we shall have a true intersection. In the other cases there will in general be no intersection at all, for the planes are in parallel regular forms, but if there is one point of intersection there will be a whole straight line—and both planes are in one regular form. Again with a little effort of imagination two regular forms may be conceived in space at once. If their planes of uniform blackness are not parallel, each plane in one regular form will intersect all the planes of the other, in straight lines. Of all these however only one will be a true intersection, namely where the plane intersects the other plane of the same blackness. It may readily be seen that all those real intersections of each plane in one form with the corresponding plane in the other form, together constitute a properly shaded plane, the intersection of the two regular forms. If the planes of uniform blackness in the two forms are parallel, any one of them will not in general intersect the corresponding one in the other form. But unless also the shading of the two forms varies at the same rate in the same direction, that is unless the regular forms are parallel, at some point the shading of one form will overtake that of the other, and here a whole plane of uniform blackness from each form will coincide, and we get a plane of intersection—to which material space happens to be parallel. The intersections of straight lines and planes with a regular form may be similarly imagined.

If we conceive two intersecting lines, the true angle between their directions would not be the same as the apparent one, just as in a plane diagram the angles between straight lines are altered. The shading would however indicate the alteration. For if we took two points of equal shading in the two straight lines, and joined them, material space would be parallel to the straight line so found, and therefore by a revolution round it, the pair of intersecting straight lines might be brought into a regular form parallel to material space, when the angle would appear at its true magnitude. Hence the perpendicular from the intersection of the two straight lines on the axis round which they were twisted, would increase in length until they were in a regular form parallel to material space, and therefore the angle within which this perpendicular fell would be diminished, and the other angle increased by the twisting, in a manner which may easily be understood from the analogous case of plane projections with which we are all familiar. It is to be observed that in the revolution here contemplated no apparent revolution would take place. The figure would appear to remain in the same plane, the angles only expanding or contracting by the straight lines sliding in the plane, and the parts remote from the axis becoming more, or less, black until they assumed the shade appropriate to the regular form in which the axis lay, parallel to material space.

In the same way we may conceive the revolution of a tetrahedron about its base. If we suppose the base to be fixed in material space, the three sides ending in the vertex will all be shaded, growing either more or less black towards the vertex, if this is not in material space. If therefore it start from material space, say towards the black direction, it will grow darker and darker. Since the perpendicular from the vertex always reaches the base in the same point, and since the angles it makes with directions in the base are all equal, and being right angles are not foreshortened by the projection, the perpendicular will not appear to revolve, but only to become shorter, until the vertex apparently sinks into the plane of the base, here attaining its maximum blackness, namely so much blacker than the base as corresponds to a difference of concentration equal to the length of the perpendicular. After this the vertex will emerge at the opposite side

of the base again, still apparently moving in the same straight
line, but getting less and less black, until as it returns to its
original shade, the perpendicular regains its original length.
By a similar succession of phases, the vertex becoming lighter
in shade however, instead of darker, the revolution may be
completed, and the vertex return to its original position.

In Book II. we saw that two tetrahedra might be equal in
every respect, but yet not congruent. This phenomenon is one
of which we have every-day experience—as in the case of
'right-' and 'left-hand' boot. And it may have struck the
reader that the expression used in the enunciation of my
proposition II. 16 'If the corners are similarly situated in
space' is rather an indefinite test of right- or left-handedness to
give in what professes to be a formal text-book. But the fact
is there is no formal geometrical test to apply, the only other
appeal would be to the right and left hand of the reader, or to
some other objective thing, which could have no place in a
subjective geometry. In that geometry we were conceiving
figures in a regular form, and we now see that it is a mere
chance which way we might happen to conceive them to be
placed in that form—whether in the same or in the opposite
way to another figure equal in every respect. The same thing
in the case of plane figures is evident. No one can define a
scalene triangle as right- or left-handed, simply because it
might be put into a given plane either way up. If however
we had a set of triangles which never moved out of a plane, we
might label one right-handed, and every other triangle equal in
every respect would be either right-handed ` like it, or left-
handed. So it is with us in material space. I am in material
space and cannot get out of it. I therefore mentally label one
of my hands 'right' and the other 'left' and no confusion can
arise. But if I ever, by magic or other arts, were allowed to
move in a fourth independent direction, I might on my return
find my right hand where my left used to be—I should have to
wear my right-hand boots on the foot I used to call my left—
with the exception of the pair I wore during my mystical
journey, which would of course have turned with me. But
to return to serious reasoning, I commend the question of
right- and left-handedness to the consideration of Euclidian
geometricians, and ask them to discuss it from their point of

view, and either to frame a geometrical definition of it, or to explain why they cannot do so.

In geometry of two independent directions a point is fixed if its distances from two fixed points remain constant. This may be expressed by saying that a triangle is the simplest rigid frame. Similarly in geometry of three independent directions a tetrahedron is the simplest rigid frame, and in geometry of four, a pentangular body. In a pentangular body there are five corners, and a side connects every two, that is in all $\dfrac{5 \cdot 4}{1 \cdot 2} = 10$ sides, a plane face connects every three, that is $\dfrac{5 \cdot 4 \cdot 3}{1 \cdot 2 \cdot 3} = 10$ faces, and a regular form connects every four, that is $\dfrac{5 \cdot 4 \cdot 3 \cdot 2}{1 \cdot 2 \cdot 3 \cdot 4} = 5$ regular forms. This method is general, and so we can calculate the corresponding figures for a rigid frame of any number of dimensions.

A point moving in a constant direction generates a straight line, this straight line moving parallel to itself in a second direction, a plane. In the same way a spread of any number of independent directions may be generated. But if each of these motions is limited in amount, we generate a terminated straight line, a parallelogram, a parallelepiped, and what we may call a parallel body in geometry of four independent directions. A parallel body will have twice as many corners as a parallelepiped, that is 16. At each corner four sides will meet, extending in four independent directions, that is the number of sides is $16 \times 4 \times \frac{1}{2} = 32$. Every two sides meeting at a corner determine a plane face, each face however unites four corners. Hence the number of plane faces is $16 \cdot \dfrac{4 \cdot 3}{1 \cdot 2} \cdot \dfrac{1}{4} = 24$. Every three sides meeting at a corner determine a regular form, but each regular form, or parallelepiped, unites eight corners. Hence the number of regular forms is $16 \cdot \dfrac{4 \cdot 3 \cdot 2}{1 \cdot 2 \cdot 3} \cdot \dfrac{1}{8} = 8$.

The above results are summarised in the following table, which it is not quite superfluous to give, as similar but incorrect tables have been given before.

| Number of i.d's. of dimensions of most extended body of Cartesian axes | 0 | 1 | 2 | 3 | 4 | $n$ |
|---|---|---|---|---|---|---|
| of axis planes | 0 | 0 | 1 | 3 | 6 | $\dfrac{n(n-1)}{1.2}$ |
| of axis forms | 0 | 0 | 0 | 1 | 4 | $\dfrac{n(n-1)(n-2)}{1.2.3}$ |
| of axis spaces of four i.d's. | 0 | 0 | 0 | 0 | 1 | $\dfrac{n(n-1)(n-2)(n-3)}{1.2.3.4}$ |
|   and so on | | | | | | |
| of corners of rigid frame | 1 | 2 | 3 | 4 | 5 | $(n+1)$ |
| of sides   ,, | 0 | 1 | 3 | 6 | 10 | $\dfrac{(n+1)n}{1.2}$ |
| of plane faces   ,, | 0 | 0 | 1 | 4 | 10 | $\dfrac{(n+1)n(n-1)}{1.2.3}$ |
| of regular forms  ,, | 0 | 0 | 0 | 1 | 5 | $\dfrac{(n+1)n(n-1)(n-2)}{1.2.3.4}$ |
|   and so on | | | | | | |
| of corners of parallel frame | 1 | 2 | 4 | 8 | 16 | $2^n$ |
| of sides   ,, | 0 | 1 | 4 | 12 | 32 | $n.2^{n-1}$ |
| of plane faces   ,, | 0 | 0 | 1 | 6 | 24 | $\dfrac{n(n-1)}{1.2}.2^{n-2}$ |
| of regular forms  ,, | 0 | 0 | 0 | 1 | 8 | $\dfrac{n(n-1)(n-2)}{1.2.3}.2^{n-3}$ |
|   and so on | | | | | | |

The reader will do well, if he really wishes to become familiar with geometry of four independent directions, to picture these figures, and the other more complex ones, to himself from various points of view, and to picture them as twisting from one view to another. If he commences to do this by the aid of the shading convention he will find after a little practice he will even be able to dispense with this aid, and to truly conceive a fourth independent direction, in the same way, if not with the same ease, as he conceives the third.

# CHAPTER IV.

WE may now discuss an alternative to the supposition that material space is a regular form, which the two objective facts we have accepted still leave open. When the gifted author of 'Flatland' conceived a race of beings living in a land of two dimensions, he tacitly assumed that that land was a plane:— it does not seem to have occurred to him to conceive a race living on the surface of a sphere, or to discuss the views of geometry they would be likely to entertain. If the sphere were very large compared with their powers of locomotion, they might indeed never find out that it was not a plane—for the two objective facts of their material space corresponding to those we have accepted as true about ours, would still hold good. In line land there would indeed be two possible alternatives, the line might either be a circle, or a helix. I propose then to show at once that for us too there is an alternative, if we only grant the two objective facts corresponding to my first and third subjective axioms; and I shall subsequently show that this is the *only* alternative consistent with the truth of those objective facts.

The objective facts which we have accepted determine, the first, that material space possesses the property that any portion of it is congruent with any other equally extensive portion, for any geometrical figure which exactly occupies the first portion may be made to exactly occupy the second, without change of size or shape. It is, in fact, "Ein in sich selbst congruenter Raum," that is, a self-congruent space. The second objective fact determines that that space is a form, whether a regular one or not.

We are naturally led at once to seek for a self-congruent form in the analogue of the sphere. We may define a circle as the locus, in a spread of two independent directions, of all points at a given distance from a given point in it; and it may be conceived to be generated by a point in a straight line revolving about a fixed point in itself, in such a spread. A sphere is a similar locus in a spread of three independent directions, and may be conceived to be generated by the revolution of a circle about a diameter, in such a spread. The locus of all points in a spread of four independent directions at a given distance from a given point in it may similarly be conceived to be generated by the revolution of a sphere about one of its diametral planes. For the regular form containing the sphere will, before completing a straight angle, have passed through every point in the spread of four independent directions, and, as in each situation of it the sphere contains all points in it at the given distance from the given point, or centre, the locus generated, when the revolution through a straight angle is complete, is the locus required. Moreover the locus is a form. For if a plane be conceived through the radius vector to any point in the locus, and extending in any direction perpendicular to the radius vector, the extremity of the radius vector may describe a circle in that plane, and in the locus. But in the space of four independent directions there is a whole group of three independent directions, each of which is perpendicular to the radius vector; and therefore the extremity of the radius vector may commence to move in any one of these directions. Hence we may call the locus a 'circular form,' and say it extends from every point in it in a complete group of three independent directions, each of which is perpendicular to the radius vector. In more familiar language, the radius vector is perpendicular to the tangent-regular-form at every point.

We can picture to ourselves the generation of a circular form, just as we pictured the revolution of a tetrahedron about its base. If the revolution of the sphere is about a diametral plane to which material space is parallel, each point in the sphere will merely appear to move in a straight line perpendicular to the axis plane, those on one side becoming lighter, and the others darker in shade as they appear to approach the axis plane. After reaching the axis plane, when the whole sphere will merely appear as a circular disc, with a double shading on it, becoming both lighter and darker towards the centre, the points of the sphere will appear respectively to emerge from the axis plane on opposite sides, and their shading to return to its original intensity as they approach material space again. During the intermediate stages the shape of the sphere will appear to be a more or less oblate spheroid. Thus if the sphere was not originally in a regular form parallel to material space, it would have begun by looking like an oblate spheroid, one pole of which was more, and the other less, black than the equator, which would be of a uniform shade. If then the revolution did not take place about the equatorial plane, each point in the sphere would not simply appear to move in a straight line perpendicular to the axis plane, but would appear to describe an ellipse of greater or less excentricity. If the axis plane extended in the direction of concentration, the sphere would always look like a doubly shaded circular disc, as described above, and the revolution of the sphere would merely appear like the ordinary revolution of such a disc about a diameter—though of course that diameter would be a doubly shaded line, which in reality represented an equator of the sphere. In any of these cases the same result would be produced; namely a spherical ball, doubly shaded from the circumference towards the centre, the one shading growing blacker, and the other less black, as it proceeded inwards.

It is not impossible to imagine two shaded clouds occupying the same space, and forming thus a diagram of concentration of a circular form. But we may also, at the cost of a certain loss of information, simply conceive, as a diagram of solidity, a spherical ball, of a density at the centre corresponding to the thickness of the ball, that is the length of a diameter, and at every other point in it to the length of a chord at right angles to the straight line from the centre to that point. Thus the density would vanish at the circumference of the ball, but it would not fade away by imperceptible degrees, but would end quite abruptly, since at the circumference the circular form would extend in the direction of concentration.

But to proceed with the investigation of the geometrical properties of the circular form, in continuation of Book III.

XV. *If a regular form intersect a circular form, the intersection is a sphere.* For if a perpendicular be conceived from the centre of the circular form to the regular form, all points in the intersection must be equidistant from the foot of this perpendicular, for if they were not they could not be equidistant from the centre of the circular form (III. 12). If the perpendicular is not less than the radius of the circular form there will of course be no intersection, and if it is equal there will be a contact. It will not be necessary to mention such obvious special cases in the other propositions.

XVI. *If two circular forms intersect, their intersection is a sphere.* For the centres of the two spheres and any point in the intersection form a triangle, all of whose sides are constant in length, and whose base is immoveable. And the perpendicular from the vertex, the point in the intersection, must meet this base always in the same point, and be of constant length, for any two situations of the triangle are congruent to each other (I. 14). Hence the intersection consists of all points at a fixed distance from the fixed foot of the perpendicular, and in directions from it perpendicular to the straight line joining the centres of the spheres—that is therefore, in a regular form (III. 11). Therefore the intersection is a sphere.

XVII. *If a plane intersect a circular form, the intersection is a circle.* This may be proved in the same way as proposition XV.

XVIII. *If a sphere intersect a circular form, the intersection is a circle.* For the intersection of the regular form in which the sphere is and the circular form is a sphere (III. 15). And that the intersection of this sphere and the given sphere, which are both in one regular form, is a circle, may be proved as in III. 16, unless the centres of the two spheres coincide. In this case, since by hypothesis they intersect, they must be one and the same sphere.

XIX. *If three circular forms intersect, their intersection is a circle.* For the intersection of two of them is a sphere, and the intersection of this with the third a circle (III. 18).

XX. *If a straight line intersect a circular form, it does so in two, and only two, points.* For the intersection consists of points equidistant from the foot of the perpendicular from the centre on the straight line. If this perpendicular is equal in length to the radius of the circular form there is of course only one point common to the straight line and form—but in this case the straight line does not intersect, but only touches it.

If a regular form through the centre of a circular form intersect it, the radius of the sphere of intersection is obviously the same as that of the form. In all other cases it is less. Hence such a sphere may be called a great sphere in the circular form. In the same way the intersection of a plane through the centre with a circular form may be called a great circle, and the points of intersection of a straight line through the centre, two poles. The intersection of two great spheres is a great circle, and the intersection of two great circles, or three great spheres (if they intersect at all), is a pair of poles. Through two great circles which intersect, or any three points which are not in the same great circle, a great sphere may be conceived. Thus the great circles joining three points which are not in one great circle form a spherical triangle, the radius of whose sphere is that of the circular form. Hence it may be shown as in spherical trigonometry that if $A$, $B$, and $C$ be the angles of the triangle, and $\Delta$ its spherical area, and $\rho$ the radius of the circular form,—

$$A + B + C - \pi = \frac{\Delta}{\rho^2}.$$

Now we have seen that a circular form may be conceived to be generated by the revolution of a sphere round any diametral plane. The whole circular form may therefore revolve in any way about its centre without ever occupying any new positions. Hence any figure in it may be carried from any one part to any other part of a spread of this form, without undergoing any change of size or shape. Moreover such a figure may revolve about any diameter of the circular form in any way, that is, remaining in the circular form it may revolve about any point in it, that is a pole of such a diameter. Similarly the figure may revolve about any diametral plane of the circular form, that is, remaining in the circular form, it may revolve about any great circle in it. It follows therefore that a space of the shape of a circular form is a self congruent space of three dimensions, and as far as the two objective facts we have accepted as true go, there is nothing to distinguish material space from such a space. If it were such a space it would follow that the things we call straight lines were really great circles, and the things we call planes great spheres of circular space. Let us consider whether we really know anything to preclude this possibility.

What do we actually know about the things we call straight lines? If I say that the edge of a certain ruler is 'straight' I may *mean* that it extends in the same direction from every point in it, but if my assertion is challenged I have to fall back upon some other test of straightness. There are four ways in which the straightness of a line may be tested, and upon each of them a geometrical definition of the term 'straight line' has been founded by one or other writer on the subject. The four tests are these:

1. Take three approximate straight edges, and fit them against each other two and two, noticing where they touch. By scraping down the places where they touch each may ultimately be made to coincide exactly with the other two all along. In practice this method is used with plane surfaces, not straight edges, and straight edges are got by the intersection of two such surfaces. This may be called the Whitworth test of straightness or flatness, for it was Sir Joseph Whitworth who first made practical use of it. It corresponds to the definitions 'Two straight lines, or planes, cannot (in general) enclose a space.'

2. Take a rigid rod, supposed to be straight, and twist it about two fixed points in itself (the rod may be conceived as a thin right circular cylinder—and in the limit, as a line). If straight, it will not change its position. Thus a turned wooden ruler of uniform thickness, is 'straight.' The corresponding definition may be stated thus—'A straight line is such that it may be twisted about two fixed points in itself without change of position.'

3. Take a fine light flexible thread and stretch it tightly between two points. The 'straight' line thus produced (the sag of the string owing to its weight being neglected) is very frequently used, by bricklayers, gardeners, and so on. It corresponds to Legendre's definition, which may be paraphrased, in order not to use terms in a different sense from what I have given to them in this book, thus 'A straight line is the shortest path between two positions.'

4. But perhaps the commonest test of all in every-day

life is that a straight line is the path of a ray of light—in an uniform medium, that is in one in which it is propagated with the same velocity in all directions. Hence it follows that, whether on the wave theory of light or on the emission theory, the locus of a given phase of a given wave, or of particles which started at a given moment, as the case may be, is the locus of a number of points whose distances from the origin of the disturbance are equal. The path of a 'ray' of light is an orthogonal trajectory of a series of such loci, and a straight line may therefore be defined as such a trajectory.

I believe one or other of these four tests corresponds to every definition of a straight line which has been made use of in geometry, except the definition by direction. Euclid's definition may perhaps be excepted, as having no meaning whatever, but he uses the first test of straightness above cited. Plato said ' A straight line is that of which the extremity hides all the rest, the eye being placed in the continuation of the line.' This is of course the fourth test. Legendre, and the majority of foreign geometricians I believe, have adopted the third test. One author only, as far as I have discovered[1], makes any theoretical use of the second test, deducing it from the idea of the locus of the points of contact of two spheres with given centres. Thus the test approximates very closely to the fourth test. The same author defines a plane in an analogous manner, which is equivalent to defining it as the locus of all points at equal distances from two given points.

Now it would be easy to show that *in a regular form* all these tests would be fair tests of the straightness of a line. But it is not difficult to show that if space is a circular form all the tests are equally applicable, but that all give, not a straight line but, a great circle.

For (1) through three points which are not in a straight line one plane only can be drawn. Hence through two points not in a diameter of a circular form, and the centre, one plane only can be drawn, and therefore through two such points in a circular form only one great circle can pass. We have therefore only to assume that material space is so large a circular form that we have never attempted to draw more than one 'straight' line through two points at opposite ends of a dia-

---

[1] 'Geometry without Axioms,' T. P. Thomson.

meter, and it would seem to us that two 'straight' lines, (though they were really great circles) could not enclose a space.

(2)   As a diametral plane of a circular form can be twisted about itself without any change of position, so can a great circle in the circular form.   The way Mr Thomson deduces his definition from the idea of a series of spheres touching each other may also be followed out in a circular form.   For a sphere in a circular form is the intersection of that circular form with a regular form.   Its centre is the foot of the perpendicular from the centre of the circular form upon this regular form, and is therefore not in the circular form at all.   The apparent centre is a point in the circular form in the diameter through the true centre, and the apparent radii are arcs of great circles through this apparent centre.   Let $A$, $B$ therefore be the apparent centres of two spheres in a circular form whose centre is $O$, and let $P$, $Q$ be their true centres, and let the two spheres touch each other at $C$, in the circular form.   Then the regular forms containing the spheres intersect in a plane through $C$.   And, since both spheres touch this plane, the directions $PC$, $QC$ are perpendicular to it.   But since the circular form extends from $C$ in the same directions as this plane does, the direction $OC$ is also perpendicular to it.   Now since the directions $OP$, $OQ$ are perpendicular to the straight lines $CP$, $CQ$, these straight lines are not parallel.   Therefore they determine a plane, which must therefore be perpendicular to the plane of contact of the two spheres, through $C$, and which must therefore contain also $OC$, which is perpendicular to this plane.   (III. 9.)   Hence the point $C$ is in the plane $OPQ$, that is in the plane $OAB$.   Therefore it is in the great circle $AB$, and this great circle is therefore the locus of such points of contact.

(3)   That a great circle is the shortest path from one point to another in a circular form may be proved in the same way that it is proved in spherical trigonometry.   Or, since all the premises, from which the analogous proposition is proved for a regular form, apply in a circular form, if the lines in the figure are none of them as great as a semi-circumference of a great circle, the proof that a straight line is the shortest path in a regular form applies to prove that the great circle is the

shortest path in a circular form, if it is not so great as a semi-circumference. (The fact that the exterior angle of a triangle is greater than the interior and opposite is proved in Euclid's manner, supplemented in the way I have indicated in Part I.)

(4) I have already shown that the true centre of a sphere in a circular form is not in that form, but in the radius vector to the apparent centre, or pole, of the sphere. The centres of a series of increasing spheres with the same apparent centre will therefore be a series of points in this radius vector, approaching the centre of the circular form. Any one of these spheres, from any point in it, extends in directions which are perpendicular both to its own radius vector from its true centre, and to the radius vector from the centre of the circular form. That is to say, it extends in directions perpendicular to a plane containing its own true centre, the centre of the circular form, and the point in question. This plane also contains its apparent centre, and is the diametral plane, whose intersection with the circular form constitutes the great circle, which is the apparent radius vector to the given point. The sphere therefore extends from the given point in directions to which the directions in which this apparent radius vector extends, are perpendicular. Thus, such a great circle, through the apparent centre of the series of (apparently) concentric spheres, is an orthogonal trajectory of the series; and so fulfils the last test we have of 'straightness.'

Thus the two objective facts we have accepted as true are not enough to prove that material space is not a circular form. Moreover if it was a circular form of large radius, we might easily not find it out without having first logically investigated the possibility of its being one, and its consequences. This will be evident if we merely consider how many generations of men lived and died without ever finding out that the surface of the earth is a sphere, and not a plane. And yet they had far more obvious means at hand wherewith to deal with the problem, for their 'straight' lines were at least straight enough to detect the curvature of the surface of the earth.

The method to be employed in this case will however doubtless have occurred to the reader at once. It is by measuring the angles of triangles, for excess. The discussion of this method is however better postponed until we see that its results would really be conclusive. To do this we must investigate whether there are no other forms which are self-congruent, besides the regular and circular forms; and this when we conceive the possibility of the variation of positions not only in four, but in any number of independent directions.

It would be difficult to conduct this investigation by the geometrical methods I have hitherto employed, so I embrace the opportunity of showing how readily the conception of direction, as I have advanced it, adapts itself to the development of the theory of analytical geometry; and I shall outline that theory from the very beginning in order to show that its methods are perfectly general, and that it is by a pure *mathematical* induction, not by mere reasoning from particular to general assertions, that its methods are applied to geometry of any number of independent directions.

# CHAPTER V.

The primary object of analytical geometry is the numerical representation of differences of position and direction. For we cannot even conceive such a thing as an absolutely fixed point, we can only assume that a certain point is fixed, that is, that it does not change its position, and determine the positions of all other points by reference to it. So also, although the common acceptance of Newton's first law of motion seems to make it probable that we do conceive of absolutely fixed directions, we have no geometrical test for fixity of direction, and so, geometrically, directions also can only be determined relatively. The problem of analytical geometry is then to determine positions and directions with reference to a given point, whose position is assumed to be unalterable, and one or more given directions.

The most elementary conception of geometry is that of a single position, of a single, fixed, point. Such a point, the datum from which we start, is called the 'Origin.' If besides this we conceive any other points, they lie in one or more directions from the origin. And having once conceived a direction, we may conceive positions varying in this direction from any position we have already conceived.

Let us then start with an origin $A$, and a second position $B$. Then we have conceived the direction $AB$. Let $C$ therefore be another point in this direction from $B$. Since we have also conceived the direction $BA$ we may also conceive points in this direction from $A$, that is we may conceive any point in the unterminated straight line $AB$.

If now the distance $\overline{AB}$ be represented by unity and the distance $\overline{AC}$ by 3, say, the positions of $B$ and $C$ with reference to $A$ as origin may be represented by the numbers 1 and 3 respectively. But now suppose we take $B$ as origin, then the position of $C$ will be represented by the distance $\overline{BC}$, that is by $(\overline{AC} - \overline{AB})$ or 2. If we apply the same rule to the case of the position $A$, since its position was formerly represented by 0, its position will now be represented by $(0-1)$, or $-1$. Hence, if we represent positions on one side of the origin by positive numbers, or coordinates as they are called, and those on the opposite side by negative numbers, a single number, with its proper sign, will represent any position. And we obtain the following universal rule :—

*If the origin is moved to a position whose old coordinate was* a, *a position whose coordinate was formerly* x *will now be represented by* (x − a).

Thus we can analytically determine any position, in a given direction or the opposite, from a given origin $O$, by a single (positive or negative) coordinate, $x$. If we now conceive another position besides all these, it lies in a new direction from $O$, and we may conceive a similar series of positions in this new direction, each determined by a single coordinate $y$, from the origin $O$, or from any of the positions formerly determined by a coordinate $x$ from $O$. Thus we have a system of positions determined by two measurements in independent directions from $O$ ; $x$ and $y$. By the first corollary to proposition I. 23 the order in which these measurements are made is indifferent. And as they are quite independent measurements the rule about change in the coordinate consequent on changing the origin, given above, will apply to each separately. Thus if the origin is moved to a position whose old coordinates were $(a, b)$ the new coordinates of $(x, y)$ are $(x - a)$, $(y - b)$.

If we now consider yet another position not included among these, we conceive a third independent direction, and to determine a position we require three coordinates. The rule for change of origin applies as before to each separately. And so we may go on *ad libitum.* Hence in a space of any number of independent directions, if the origin is moved to a point $(abc...)$ and if $(xyz...)$ be the old and $(x'y'z'...)$ the new coordinates of any point, we have

$$x' = x - a; \quad y' = y - b; \quad z' = z - c; ...$$

and $\qquad x = x' + a; \quad y = y' + b; \quad z = z' + c; ... \Bigg\} \quad \text{...... (1).}$

Straight lines through the origin in the given independent

directions in which the coordinates are measured are called the coordinate axes. Every two of these determine a plane, called an axis plane. Every three determine a regular form, called an axis form. And so on. Thus in geometry of four independent directions there are four axes, six axis planes, and four axis forms.

The obvious way of indicating a direction analytically is by giving the coordinates of a point at unit distance from the origin. It follows from the principles of similar triangles established by Euclid in his sixth book, that if $(xyz...)$ be the coordinates of any point at a distance $s$ from the origin, its direction from the origin is represented by $\left(\dfrac{x}{s}, \dfrac{y}{s}, \dfrac{z}{s}...\right)$.

To find its direction from any other point $(x'y'z'...)$ whose distance from it is $s'$ we have only to move the origin to the point $(x'y'z'...)$, so that the coordinates of $(xyz...)$ become $(x-x')(y-y')(z-z')...$; and therefore its direction from $(x'y'z'...)$ is represented by $\left(\dfrac{x-x'}{s'}, \dfrac{y-y'}{s'}, \dfrac{z-z'}{s'}...\right)$. Similarly, to find the direction of the point $(x'y'z'...)$ from $(xyz...)$ we move the origin to $(xyz...)$; and as $s'$, the distance between the points, remains the same, the direction is represented by $\left(\dfrac{x'-x}{s'}, \dfrac{y'-y}{s'}, \dfrac{z'-z}{s'}...\right)$, that is, by the same numbers, or direction coefficients as we may call them, as before, but with their signs reversed.

To find an expression for the distance of any point from the origin, let $OL, LM, MN,...$ be the coordinates $(xyz...)$ of any position $P$ at a distance $s$ from the origin $O$. Let $(\alpha\beta\gamma...)$ be the angles which $OP$ makes with the directions of the axes, and conceive perpendiculars $PA, PB, PC...$ from $P$ to each of the axes.

Then if each of the points $L, M, N,...$ be projected upon $OP$ by straight lines perpendicular to it, it is obvious that $OP$ is equal to the sum of these projections, that is that

$$s = x \cos \alpha + y \cos \beta + z \cos \gamma + \ldots\ldots$$

But $\quad \cos \alpha = \dfrac{OA}{OP} = \dfrac{OA}{s}$. Hence

$$s^2 = x \cdot OA + y \cdot OB + z \cdot OC + \ldots\ldots,$$

and $OA$ is equal to the sum of $OL$ and $LA$, that is to the sum of $x$, and the sum of the projections of $y$, $z$... upon the axis of $x$, by straight lines perpendicular to it. That is, if $(xy)$ represent the angle between the directions of $x$ and $y$, and so on

$$s^2 = x \cdot x + x \cdot y \cos(xy) + x \cdot z \cos(xz) + \ldots\ldots$$

$$+ y \cdot x \cos(yx) + y \cdot y + y \cdot z \cos(yz) + \ldots\ldots$$

$$+ z \cdot x \cos(zx) + zy \cos(zy) + z \cdot z + \ldots\ldots$$

$$\ldots\ldots\ldots\ldots\ldots\ldots\ldots\ldots\ldots\text{and so on}$$

$$= x^2 + y^2 + z^2 + \ldots\ldots + 2xy \cos(xy) + 2yz \cos(yz) + \ldots\ldots (2),$$

there being $p$ terms of the first kind, and $\dfrac{p \cdot p - 1}{2\,!}$ of the second.

If $(lmn\ldots)$ be the direction coefficients of the direction $OP$, dividing the above equation by $s^2$ we get, since $l = \dfrac{x}{s}$, $m = \dfrac{y}{s}$ and so on,

$$l^2 + m^2 + n^2 + \ldots + 2lm \cos(xy) + 2mn \cos(yz) + \ldots = 1 \ldots\ldots(3).$$

This equation of condition subsists then for all direction coefficients.

The distance of $(xyz\ldots)$ from any other point $(x'y'z'\ldots)$ can be deduced at once from the above formulae by moving the origin to $(x'y'z'\ldots)$, that is by writing $(x - x')$ for $x$, $(y - y')$ for $y$ and so on in the formula (2). If the axes are all at right angles to each other the cosines of the angles between them are all zero and so equations (2) and (3) become

$$s^2 = x^2 + y^2 + z^2 + \ldots\ldots, \qquad l^2 + m^2 + n^2 + \ldots\ldots = 1.$$

The condition that any number of directions $12\ldots q$ shall not be independent of each other in a space of $p$ independent directions, may be found from the consideration that if they are not, after moving a distance $s_1$, from the origin in any one of the given directions it must be possible to return to it by movements (positive or negative) $s_2\,s_3\ldots s_q$ in the remaining directions. Hence given any value of $s_1$ it must be possible to find values of $s_2\,s_3\ldots s_q$ to satisfy the $p$ equations

$$\left.\begin{array}{l} s_1 l_1 + s_2 l_2 + s_3 l_3 + \ldots\ldots s_q l_q = 0 \\ s_1 m_1 + s_2 m_2 + s_3 m_3 + \ldots\ldots s_q m_q = 0 \\ \ldots\ldots\ldots\ldots\ldots\ldots \text{ and so on} \end{array}\right\}.$$

We have then to eliminate $(q - 1)$ quantities from $p$ equations.

If we can do so, we may divide by $s_1$, the remaining quantity, and have left one or more equations of condition that the directions shall not be independent. Thus if $q$ is greater than $p$ these equations will contain some of the variables $s_2\ s_3...$ and so a solution can always be found, i.e. the directions are not independent. If $q$ is equal to or less than $p$ we obtain one or more conditions, in the form of determinants equated to zero. As such symmetrical determinants occur frequently in the next few pages, I shall use the symbol $\phi\,\dfrac{lmn......}{123......p}$ for a determinant formed of columns of $l$'s, $m$'s and so on, the rows being distinguished by the suffixes $123......p$. If the number of letters in the numerator is less than the number of suffixes, the number is to be made up with columns of units.

The condition then that $p$ directions be not independent is

$$\phi\,\frac{lmn......}{123......p} = 0,$$

where all the $p$ letters appear in the numerator. The conditions that any less number, $q$, be not independent are of the same form, but only $q$ of the $p$ letters appearing in each numerator and of numbers in each denominator. Hence if $p - q = r$, the number of conditions may be written either

$$\frac{p\,.\,p-1\,.\,p-2......p-r+1}{r\,!} \quad \text{or} \quad \frac{p\,.\,p-1\,.\,p-2......p-q+1}{q\,!},$$

but of this number only $(r + 1)$ are independent, for they are all deduced from $p$ equations after eliminating $(q - 1)$ quantities.

The condition that two directions may be at right angles is easily found. For if the directions 1 and 2 are at right angles the distances of a point $(l_1 m_1 n_1...)$ from two points $(l_2 m_2 n_2...)$ $(-l_2, -m_2, -n_2,...)$ each at unit distance from the origin, must be equal. Hence

$$(l_1 - l_2)^2 + (m_1 - m_2)^2 + (n_1 - n_2)^2 + ......$$
$$+ 2\,(l_1 - l_2)\,(m_1 - m_2)\cos(xy) + ......$$
$$= (l_1 + l_2)^2 + (m_1 + m_2)^2 + (n_1 + n_2)^2 + ......$$
$$+ 2\,(l_1 + l_2)\,(m_1 + m_2)\cos(xy) + ......$$

From which we deduce the condition

$$l_1 l_2 + m_1 m_2 + n_1 n_2 + ... + (l_1 m_2 + l_2 m_1)\cos(xy) + ... = 0... \quad (5).$$

In the case of rectangular axes this reduces to

$$l_1 l_2 + m_1 m_2 + n_1 n_2 + \ldots\ldots = 0.$$

The condition that two directions may be identical or opposite may be put in a similar form, namely

$$l_1 l_2 + m_1 m_2 + n_1 n_2 + \ldots\ldots = \pm 1 \ldots\ldots\ldots\ldots(6).$$

For if to twice the above equation (6) with the lower sign we add the equations

$$l_1^2 + m_1^2 + n_1^2 + \ldots\ldots = 1,$$

$$l_2^2 + m_2^2 + n_2^2 + \ldots\ldots = 1,$$

or if from these we subtract twice equation 6 with the upper sign, we get

$$(l_1 \pm l_2)^2 + (m_1 \pm m_2)^2 + (n_1 \pm n_2)^2 + \ldots\ldots \&c. = 0,$$

which requires that each of the expressions in the brackets should vanish.

It is easy in a similar manner to find the trigonometrical ratios of the angle between any two directions. For the distance between the points $(l_1 m_1 n_1)$ and $(l_2 m_2 n_2)$, both being at unit distance from the origin, is the chord of this angle. Hence if the axes are rectangular we have

$$\sin \frac{A}{2} = \tfrac{1}{2} \sqrt{(l_1 - l_2)^2 + (m_1 - m_2)^2 + (n_1 - n_2)^2 + \ldots\ldots \&c.}$$

$$= \sqrt{\tfrac{1}{2}(1 - l_1 l_2 - m_1 m_2 - n_1 n_2 - \ldots\ldots \&c.)},$$

$$\cos \frac{A}{2} = \sqrt{1 - \sin^2 \frac{A}{2}}$$

$$= \sqrt{\tfrac{1}{2}(1 + l_1 l_2 + m_1 m_2 + n_1 n_2 + \ldots\ldots \&c.)},$$

$$\cos A = \cos^2 \frac{A}{2} - \sin^2 \frac{A}{2} = l_1 l_2 + m_1 m_2 + n_1 n_2 + \ldots\ldots \&c.,$$

$$\sin A = \sqrt{1 - (l_1 l_2 + m_1 m_2 + n_1 n_2 + \ldots\ldots \&c.)^2};$$

and the latter expression may easily be shown to be equal to

$$\left\{ \left(\phi \frac{lm}{12}\right)^2 + \left(\phi \frac{mn}{12}\right)^2 + \ldots\ldots \&c. \right\}^{\frac{1}{2}},$$

the bracket containing determinants formed from all the combinations of the coefficients two at a time. The conditions (5) and (6) may therefore be replaced by equating this expression to 1 and 0 respectively. Similar expressions for oblique axes may be found by exactly the same methods.

We have seen that the directions of the axes must be independent directions, but with this restriction it is still possible to twist them into various directions. Any possible alterations in the directions of the axes may be performed by successive twists of two axes at a time in their own axis plane, and each of such twistings will only affect the coordinates parallel to the axes twisted, though in the case of oblique axes it will also affect the angles between these axes and the rest. Suppose, for example, in a space of four independent directions it is required to twist the axes into four new independent directions 1, 2, 3, 4 that is, until they pass through the four points $(l_1 m_1 n_1 k_1)$ $(l_2 m_2 n_2 k_2)$ and so on, all at unit distance from the origin. This can be done in six stages, thus—

| A rotation in the plane of | brings the axis of $x$ to | that of $y$ to | that of $z$ to | and that of $w$ to |
|---|---|---|---|---|
| $(xy)$ | $l_1 m_1 0 0$ | $l_2 m_2 0 0$ | $0\,0\,0\,0$ | $0\,0\,0\,0$ |
| $(yz)$ | $l_1 m_1 0 0$ | $l_2 m_2 n_2 0$ | $0\,m_3 n_3 0$ | $0\,0\,0\,0$ |
| $(zw)$ | $l_1 m_1 0 0$ | $l_2 m_2 n_2 0$ | $0\,m_3 n_3 k_3$ | $0\,0\,m_4 k_4$ |
| $(wx)$ | $l_1 m_1 0 k_1$ | $l_2 m_2 n_2 0$ | $0\,m_3 n_3 k_3$ | $l_4 0\,m_4 k_4$ |
| $(xz)$ | $l_1 m_1 n_1 k_1$ | $l_2 m_2 n_2 0$ | $l_3 m_3 n_3 k_3$ | $l_4 0\,m_4 k_4$ |
| $(wy)$ | $l_1 m_1 n_1 k_1$ | $l_2 m_2 n_2 k_2$ | $l_3 m_3 n_3 k_3$ | $l_4 m_4 n_4 k_4$ |

The coordinates in the table being all referred to the original axes. This method is sometimes convenient. Or else general formulae may be obtained thus,—

We observe that if a point can move from the origin to a given position by a movement $s$, in a direction $(lmn......)$, it can also move to it by movements $ls, ms, ns,......$ in the directions of the axes of $(x, y, z......)$ respectively, these being the coordinates of the position.

Let it then be required to find the coefficients $(\lambda'\mu'\nu'......)$ of a direction referred to a set of axes $(x'y'z'......)$, whose coefficients referred to a given set of axes $(xyz......)$ were $(\lambda\mu\nu......)$, the directions of the new axes referred to the old being $(l_1 m_1 n_1......)$ $(l_2 m_2 n_2......)$ and so on.

Then by hypothesis a point can move from the origin to

the point $P$, at unit distance from it, whose old coordinates are $(\lambda\mu\nu\ldots\ldots)$, either

(1)　by movements $(\lambda\mu\nu\ldots\ldots)$ in the directions of the axes of $(x,\ y,\ z\ldots\ldots)$ respectively, or

(2)　by movements $(\lambda'\mu'\nu'\ldots\ldots)$ in the directions of the axes of $(x'y'z'\ldots\ldots)$ respectively.

But these latter movements may be replaced as follows—

$\lambda'$ by $(l_1\lambda')$, $(m_1\lambda')$, $(n_1\lambda')$ in the directions $(xyz\ldots\ldots)$ respectively [for $(\lambda'00\ldots\ldots)$ is a point at distance $\lambda'$ from the origin in the direction $(l_1m_1n_1\ldots\ldots)$],

$\mu'$ by $(l_2\mu')$, $(m_2\mu')$, $(n_2\mu')$ ... in the same directions, and so on.

Or, as the movements may, by (I. 23, cor. (ii)) be taken in any order, a point may be moved from the origin to $P$ by movements

$$(l_1\lambda' + l_2\mu' + l_3\nu' + \ldots\ldots)\ \text{in the direction of } x\,;$$
$$(m_1\lambda' + m_2\mu' + m_3\nu' + \ldots\ldots)\ \text{in the direction of } y\,;$$
$$(n_1\lambda' + n_2\mu' + n_3\nu' + \ldots\ldots)\ \text{in the direction of } z\,;$$
$$\ldots\ldots\ldots\ldots\ldots\ldots\ldots\ldots\ldots\ \text{and so on.}$$

These therefore are the coordinates of $P$ referred to the axes $(xyz\ldots\ldots)$. That is to say,

$$\left.\begin{aligned}
\lambda &= l_1\lambda' + l_2\mu' + l_3\nu' + \ldots\ldots \\
\mu &= m_1\lambda' + m_2\mu' + m_3\nu' + \ldots\ldots \\
\nu &= n_1\lambda' + n_2\mu' + n_3\nu' + \ldots\ldots \\
&\ldots\ldots \text{ and so on}
\end{aligned}\right\}\ \ldots\ldots\ldots\ldots(7).$$

Whence we get also—

$$\left.\begin{aligned}
\lambda' &= \frac{\begin{vmatrix} \lambda & \mu & \nu & \ldots \\ l_2 & m_2 & n_2 & \ldots \\ l_3 & m_3 & n_3 & \ldots \\ \ldots\ldots\ldots\ldots \end{vmatrix}}{\phi\dfrac{lmn\ldots}{123\ldots}} = \frac{\lambda\phi\dfrac{mn\ldots}{23\ldots} + (-)^{p-1}\mu\phi\dfrac{ln\ldots}{23\ldots} + \&\text{c.}}{\phi\dfrac{lmn\ldots}{123\ldots}} \\[2em]
(-)^{p-1}\mu' &= \frac{\begin{vmatrix} \lambda & \mu & \nu & \ldots \\ l_1 & m_1 & n_1 & \ldots \\ l_3 & m_3 & n_3 & \ldots \\ \ldots\ldots\ldots\ldots \end{vmatrix}}{\phi\dfrac{lmn\ldots}{123\ldots}} = \frac{\lambda\phi\dfrac{mn\ldots}{13\ldots} + (-)^{p-1}\mu\phi\dfrac{ln\ldots}{13\ldots} + \&\text{c.}}{\phi\dfrac{lmn\ldots}{123\ldots}}
\end{aligned}\right\}\ (8),$$

and so on; where, in the determinants in the numerators of the second expressions, one number, namely that corresponding to the coefficient ($\lambda'$) on the left of the equation, and one letter, namely that corresponding to the coefficient ($\lambda$) of each determinant, is omitted, leaving determinants of the order ($p-1$).

To deduce from these formulae the new coordinates ($x'y'z'......$) of a position whose old coordinates were ($xyz......$) we observe that the distance of the position from the origin $s$, is unaltered; and its direction from the origin, referred to the old coordinates, is $\left(\dfrac{x}{s}, \dfrac{y}{s}, \dfrac{z}{s} ......\right)$. Substituting these in (8) we get the values of the coefficients of the same direction referred to the new axes, that is, of $\dfrac{x'}{s}, \dfrac{y'}{s}$, and so on. As the equations are all homogeneous the $s$ multiplies out, and the result is the same as if we had written $x$ for $\lambda$, $x'$ for $\lambda'$ and so on, throughout equations (7) and (8).

If only two axes are rotated in their own plane the formulae (7) and (8) reduce to

$$\lambda = l_1\lambda' + l_2\mu' \atop \mu = m_1\lambda' + m_2\mu' \} \text{ and } \left. \begin{array}{l} \lambda' = \dfrac{1}{\phi\dfrac{lm}{12}}(m_2\lambda - l_2\mu) \\[3mm] -\mu' = \dfrac{1}{\phi\dfrac{lm}{12}}(m_1\lambda - l_1\mu) \end{array} \right\} ...(9).$$

In the special case where both the old and new systems of axes are rectangular, equations (8) may be much simplified. This may be proved by the method of successive rotations of two axes at a time, referred to above. In this case, since the direction at right angles to ($l, m$) in the plane of ($xy$) is the direction ($-m, l$), equations 9 become

$$\lambda = l\lambda' - m\mu' \atop \mu = m\lambda' + l\mu' \} \text{ and } \lambda' = l\lambda + m\mu \atop \mu' = -m\lambda + l\mu \}.........(10).$$

for, $\phi\dfrac{lm}{12} = l^2 + m^2 = 1$.

Hence the effect on the coordinates ($xyz......$) of any position, of a twist of the axes of ($xy$) in their own plane till they pass through the points ($l_1 m_1 0 0......$) ($l_2 m_2 0 0......$) that is, till the axis of $x$ extends in the direction

$$\left(\frac{l_1}{\sqrt{l_1^2 + m_1^2}}, \frac{m_1}{\sqrt{l_1^2 + m_1^2}}, 0, 0......\right)$$

is that they become respectively

$$\frac{l_1 x + m_1 y}{\sqrt{l_1^2 + m_1^2}}; \quad \frac{-m_1 x + l_1 y}{\sqrt{l_1^2 + m_1^2}}; \quad z; \quad w; \quad \ldots\ldots \&c.$$

And the new coefficients of the direction $(l_1 m_1 n_1 \ldots\ldots)$ are

$$\sqrt{l_1^2 + m_1^2}; \quad 0; \quad n_1; \quad k_1; \quad \ldots\ldots \&c.$$

Hence the next twist, in the axis plane $(xz)$, has to bring the axis of $x$ into the direction

$$\frac{\sqrt{l_1^2 + m_1^2}}{\sqrt{l_1^2 + m_1^2 + n_1^2}}; \quad 0; \quad \frac{n_1}{\sqrt{l_1^2 + m_1^2 + n_1^2}}; \quad 0; \ldots\ldots \&c.$$

Therefore the coordinates of $P$ after this twist become

$$\frac{l_1 x + m_1 y + n_1 z}{\sqrt{l_1^2 + m_1^2 + n_1^2}}; \quad \frac{-m_1 x + l_1 y}{\sqrt{l_1^2 + m_1^2}}; \quad \frac{-n_1(l_1 x + m_1 y) + (l_1^2 + m_1^2) z}{\sqrt{l_1^2 + m_1^2}\sqrt{l_1^2 + m_1^2 + n_1^2}}, \quad w$$

and the new coefficients of the direction $(l_1 m_1 n_1 \ldots\ldots)$ are

$$\sqrt{l_1^2 + m_1^2 + n_1^2}; \quad 0; \quad 0; \quad k; \quad \ldots\ldots \&c.$$

If we continue this process $(p-1)$ times, that is, if we twist the axis of $x$ with reference to each of the other axes in succession, the direction coefficients $(l_1 m_1 n_1 \ldots)$ become all zero except the first, which is unity since

$$l_1^2 + m_1^2 + n_1^2 + \ldots\ldots (p \text{ terms}) = 1,$$

and the coordinate $x'$ of the position $P$ becomes

Similarly
$$\left.\begin{aligned}
x' &= l_1 x + m_1 y + n_1 z + \ldots\ldots \\
y' &= l_2 x + m_2 y + n_2 z + \ldots\ldots \\
z' &= l_3 x + m_3 y + n_3 z + \ldots\ldots \\
&\ldots\ldots\ldots\ldots\ldots\ldots \text{ and so on}
\end{aligned}\right\} \quad \ldots\ldots\ldots\ldots(11)$$

and in these equations as before we may write $\lambda'$ for $x'$, and $\lambda$ for $x$, and so on, and so get the new direction coefficients of a direction. For example, the old coefficients of the old axis of $x$ were $(1, 0, 0, 0, \ldots\ldots)$. Thus the new coefficients of the old axis of $x$ are $(l_1 l_2 l_3 \ldots\ldots)$. Thus, considering the axes $(x' y' z' \ldots\ldots)$ as the original ones, formula (11) gives for the coordinate $x$ of a position referred to the other axes

$$x = l_1 x' + l_2 y' + l_3 z' + \ldots\ldots$$

which agrees with (7).

# CHAPTER VI.

IF $(xyz......)$ is any point in a space of $p$ independent directions what is represented by a single equation—

$$F(xyz......) = 0 .....................(1).?$$

The equation obviously represents the locus of a number of points obeying certain conditions. As (in general) the variables may vary continuously in the equation, the locus will (in general) be a spread of some kind. Let us then investigate the directions in which it extends from any point in it.

If $(x + \Delta x)(y + \Delta y)(z + \Delta z)......$ be a point in the locus near $(xyz......)$ at a distance $\Delta s$ from it, then the direction from $(xyz......)$ to it is represented by the direction coefficients

$$\frac{\Delta x}{\Delta s}, \frac{\Delta y}{\Delta s}, \frac{\Delta z}{\Delta s} ......$$

If the distance $\Delta s$ is indefinitely diminished, in the limit this direction is a direction in which the spread extends from the point $(xyz......)$, and its coefficients are

$$\frac{dx}{ds}, \frac{dy}{ds}, \frac{dz}{ds} ......$$

Now $s$ is some function of $xyz.......$ Hence differentiating (1) with respect to $s$ we get

$$\frac{dF}{dx} \cdot \frac{dx}{ds} + \frac{dF}{dy} \cdot \frac{dy}{ds} + \frac{dF}{dz} \cdot \frac{dz}{ds} + ...... = 0 .........(2).$$

The quantities $\frac{dF}{dx}$, are constant at any given point in the locus, and may be replaced by letters $XYZ......$ which are understood to be dependent on $xyz......$, the coordinates of the given point. The quantities $\frac{dx}{ds}$ &c. are the direction coefficients of any direction in which the spread extends from this given point, and may be replaced by $(\lambda\mu\nu\ldots\ldots\omega)$. Thus the equation

$$X\lambda + Y\mu + Z\nu + \ldots\ldots = 0 \ldots\ldots\ldots\ldots(3),$$

determines in what directions the spread extends.

Now the $p$ quantities $(\lambda\mu\nu\ldots\ldots\omega)$ are not quite independent, for as we have seen, being direction coefficients, they are connected by an equation of condition which we found in the last chapter (formula 3). This condition also prevents their all being zero. Hence we may consider one of them $\lambda$, to be determined by this condition in terms of the others, and not to be zero, we may therefore divide equation (3) by $\lambda$ and have left an equation of the first degree in $(p-1)$ arbitrary variables,

$$\frac{\mu}{\lambda}, \frac{\nu}{\lambda}\ldots\ldots\frac{\omega}{\lambda}.$$

Now in equation (3) any one of the constant quantities $XY\ldots\ldots$ will be zero for every point in the locus, if the corresponding coordinate does not appear in equation (1). But even if it does appear, the quantity may be zero for a particular point in the locus, for it is got by differentiating $F$ with respect to one of the variables in it, and then giving these variables particular values, which may of course make the result vanish. But suppose that at a given point $(x'y'z'\ldots\ldots)$ in the locus, $q$ out of the $p$ quantities $X$, $Y$, $\ldots\ldots$ do not vanish, including $X$. Then (3) becomes an equation in $(q-1)$ arbitrary ratios $\frac{\mu}{\lambda}, \frac{\nu}{\lambda}, \ldots\ldots \frac{\pi}{\lambda}$. We may therefore assign any values we please to all but one $\left(\text{say } \frac{\mu}{\lambda}\right)$ of these ratios and also to all the $(p-q)$ coefficients which do not occur in the equation, and we may then determine $\frac{\mu}{\lambda}$. Suppose we ascribe the value zero to each one of these arbitrary ratios. We have then found one direction in which the spread extends, which we may denote by $(\lambda_1\mu_1 00 \ldots\ldots)$.

Now vary each of the $(p-2)$ arbitrary values we have assumed in turn, making each in turn unity instead of zero for instance. In the case of the $(q-2)$ arbitrary ratios which appear in the equation (3) a redetermination of $\frac{\mu}{\lambda}$ will be necessary;

but in the case of the $(p-q)$ which do not appear in the equation this will not be necessary. We shall now have $(p-1)$ directions in which the spread extends from $(x'y'z' \ldots\ldots)$. It only remains to show that they are independent directions, and that all other directions in which the spread extends are dependent upon them.

Now the complete determinant formed from these $(p-1)$ direction coefficients $(\lambda_1\nu_1 \ldots\ldots \omega_1)$ $(\lambda_2\nu_2 \ldots\ldots \omega_2) \ldots\ldots$ omitting any one of the coefficients, $\mu_1\mu_2 \ldots\ldots$ from each, may be developed thus—

$$\phi \frac{\lambda\nu \ldots\ldots \omega}{12\ldots\ldots(p-1)}$$

$$= \lambda_1\phi \frac{\nu \ldots\ldots \omega}{23 \ldots\ldots (p-1)} + (-)^p\lambda_2\phi \frac{\nu \ldots\ldots \omega}{34 \ldots\ldots (p-1).1}$$

$$+ \ldots\ldots \pm \lambda_{(p-1)}\phi \frac{\nu \ldots\ldots \omega}{12 \ldots\ldots (p-2)}.$$

Now none of the quantities $\lambda_1$, $\lambda_2 \ldots\ldots$ are zero, and of each of the quantities $(\nu \ldots\ldots \omega)$ in each direction after the first, all are zero except one, in each direction, whereas in the first direction all are zero. Hence each determinant on the right of the above equation has a complete row of zeros (that corresponding to direction 1) except the first, the letters in which are all zero except those on the leading diagonal. Therefore this term on the right hand of the equation remains alone, and is not zero. Therefore the $(p-1)$ directions are independent of each other.

Now let us select any other direction in which the spread extends, and which therefore satisfies equation (3). We thus have $p$ directions in all. If the $q$ coefficients of this new direction which appear in the equation (3) are equal or proportional to the first $q$ of any of the former $(p-2)$ directions, then the determinant

$$\phi \frac{\lambda\mu\nu \ldots\ldots \pi}{123 \ldots\ldots q}$$

formed from these first $q$ coefficients will have two rows equal or proportional, and will therefore vanish. But if not we obtain one more independent and consistent equation of form (3), that is $q$ equations in all, which suffice to eliminate the $q$ quantities

$XY$...... none of which by hypothesis are zero, and so we are left with the same condition as above, namely

$$\phi \frac{\lambda\mu\nu \,\ldots\ldots\, \pi}{123 \,\ldots\ldots\, q} = 0.$$

Now if $\psi$ be one of the remaining $(p-q)$ direction coefficients we have

$$\phi \frac{\lambda\mu\nu \,\ldots\ldots\, \pi, \psi}{123 \,\ldots\ldots\, q, (q+1)}$$

$$= \psi_{q+1} \phi \frac{\lambda\mu\nu \,\ldots\ldots\, \pi}{123 \,\ldots\ldots\, q} + (-)^q \, \psi_1 \phi \frac{\lambda\mu\nu \,\ldots\ldots\, \pi}{23 \,\ldots\ldots\, q\,(q+1)} + \ldots \&\text{c.}$$

And we gave the value zero to all the ratios $\frac{\psi}{\lambda}$ except one. As $\psi$ is not among the first $q$ direction coefficients, it was one of those after the $q^{\text{th}}$ which was not zero, say the $(q+1)^{\text{th}}$. Thus all the terms on the right of the above equation after the first vanish on account of the $\psi$ coefficient, and the first vanishes because, as we have seen, its determinant $\phi$ vanishes. In the same way it can be proved that we may form a determinant with yet another of the remaining $(p-q)$ coefficients, which will vanish; and by proceeding in this way to the end, we find that

$$\phi \frac{\lambda\mu\nu \,\ldots\ldots\, \omega}{123 \,\ldots\ldots\, p} = 0.$$

That is the $p$ directions are not independent, and therefore the spread represented by equation (1) extends from every point in it in $(p-1)$ and no more independent directions. We may call such a spread one of the $(p-1)^{\text{th}}$ order.

In the same way it may be shown that two simultaneous independent and consistent equations

$$\left.\begin{array}{l} F_1\,(xyz\,\ldots\ldots) = 0 \\ F_2\,(xyz\,\ldots\ldots) = 0 \end{array}\right\} \ldots\ldots\ldots\ldots\ldots\ldots\ldots(4),$$

in general represent a spread of the $(p-2)^{\text{th}}$ order.

For as above we may deduce the equations

$$\left.\begin{array}{l} X_1\lambda + Y_1\mu + Z_1\nu + \ldots\ldots = 0 \\ X_2\lambda + Y_2\mu + Z_2\nu + \ldots\ldots = 0 \end{array}\right\} \ldots\ldots\ldots\ldots(5).$$

And even if both these equations (5) contain all the $p$ coefficients $(\lambda\mu\nu \,\ldots\ldots\, \omega)$ we may eliminate one of them and so obtain an equation in $(p-1)$ of them, from which we may show as above that the locus extends in $(p-2)$ independent directions, for we may ascribe $(p-2)$ arbitrary values to the ratios of $(p-1)$ of the coefficients, determining the $p^{\text{th}}$ coefficient in each case from one of the equations (5). And if equations (5) are independent we may show as above that there cannot be more than $(p-2)$ independent directions which satisfy them.

Thus two simultaneous independent and consistent equations in general represent a spread of the $(p-2)^{\text{th}}$ order. Or we may say that two spreads of the $(p-1)^{\text{th}}$ order in general intersect in a spread of the order $(p-2)$.

And so generally $q$ equations represent in a space of $p$ independent directions a spread of the $(p-q)^{\text{th}}$ order, or the intersection of $q$ spreads of the order $(p-1)$ is in general a spread of the order $(p-q)$.

Thus a line will in general be represented by $(p-1)$ equations, a surface by $(p-2)$, a form by $(p-3)$ and so on.

The condition that a spread shall be a regular one is that the equation or equations determining the directions in which it extends should be the same for every point in the spread. That is, the differentials $\dfrac{dF}{dx}$, $\dfrac{dF}{dy}$ and so on, must all be constants. Hence all the equations representing regular spreads must be equations of the first degree.

The equations to a straight line may be got at in another way. Let $(abc\ \ldots\ldots)$ be any point in a straight line and $(lmn\ \ldots\ldots)$ the direction in which it extends, $s$ the distance of any other point in it from $(abc\ \ldots\ldots)$. Then we have

$$x = a + ls\ ;\ y = b + ms\ ;\ z = c + ns\ \ldots\ldots \text{ and so on.}$$

Or
$$\frac{x-a}{l} = \frac{y-b}{m} = \frac{z-c}{n} = \ldots\ldots = s \ldots\ldots\ldots (6).$$

(The last equation is of course not independent of the others.)

If this straight line passes through a second known point $(a'b'c'\ \ldots\ldots)$ at a distance $s'$ from $(abc\ \ldots\ldots)$ then

$$l = \frac{a'-a}{s'}\ ;\quad m = \frac{b'-b}{s'}$$

and so on. Hence the equations to the straight line joining two points are,

$$\frac{x-a}{a'-a} = \frac{y-b}{b'-b} = \frac{z-c}{c'-c} = \ldots\ldots = \frac{s}{s'} \ldots\ldots(7).$$

Similarly if $(abc\ldots\ldots)$ be a known point in a plane which extends in two known directions, 1 and 2, from it, then if it extend in any third direction, we have

$$\phi\frac{lmn}{123} = \phi\frac{mnk}{123} = \ldots\ldots = 0.$$

But $l_3 = \dfrac{x-a}{s}$, $m_3 = \dfrac{y-b}{s}$ $\ldots\ldots$ and so on.  Hence substituting

in the above equations and expanding,

$$\left.\begin{aligned}(x-a)\,\phi\,\frac{mn}{12} + (y-b)\,\phi\,\frac{nl}{12} + (z-c)\,\phi\,\frac{lm}{12} = 0 \\[2mm] (y-b)\,\phi\,\frac{nk}{12} + (z-c)\,\phi\,\frac{km}{12} + (w-d)\,\phi\,\frac{mn}{12} = 0 \\[2mm] \ldots\ldots \text{ and so on} \ldots\ldots\end{aligned}\right\}\ldots\ldots(8),$$

are the equations to the plane.  There will altogether be $\dfrac{p\,.\,p-1\,.\,p-2}{3\,!}$ of these equations, but only $(p-2)$ of them will be independent.

If we are given three points 1, 2, 3 in the plane which are not in a straight line, then if $s's''$ be the distances from 1 to 2 and to 3 respectively the conditions that the direction to any point $x$ from $(abc\ldots)$ shall be dependent on these directions may be written

$$\begin{vmatrix} \dfrac{x-a}{s} & \dfrac{y-b}{s} & \dfrac{z-c}{s} \\[3mm] \dfrac{a_1-a_2}{s'} & \dfrac{b_1-b_2}{s'} & \dfrac{c_1-c_2}{s'} \\[3mm] \dfrac{a_1-a_3}{s''} & \dfrac{b_1-b_3}{s''} & \dfrac{c_1-c_3}{s''} \end{vmatrix} = 0 \text{ and so on.}$$

This may be reduced to the series of equations

$$\left.\begin{aligned}(x-a)\,\phi\,\frac{bc}{123} + (y-b)\,\phi\,\frac{ca}{123} + (z-c)\,\phi\,\frac{ab}{123} = 0 \\[2mm] (y-b)\,\phi\,\frac{cd}{123} + (z-c)\,\phi\,\frac{db}{123} + (w-d)\,\phi\,\frac{bc}{123} = 0 \\[2mm] \ldots\ldots \text{ and so on} \ldots\ldots\end{aligned}\right\}\ldots\ldots(9).$$

And so we may find the equations to any regular spread. These equations all apply to oblique as well as to rectangular axes.

If the equation, referred to rectangular axes,

$$Ax + By + Cz + \ldots\ldots = K \qquad (10),$$

represent a spread of the order $(p-1)$ in a space of $p$ independent directions, then the directions in which it extends from any point are determined by the equation

$$A\lambda + B\mu + C\nu + \ldots\ldots = 0.$$

Hence a direction whose coefficients are proportional to $(A, B, C \ldots)$ is at right angles to every direction in which the spread extends, that is, it is perpendicular to the spread. If therefore $(lmn\ldots)$ be this direction, equation 10 may be written

$$lx + my + nz + \ldots\ldots = \frac{K}{\sqrt{A^2 + B^2 + C^2 + \ldots\ldots}} \ldots (11).$$

Now a straight line through the origin in this direction, perpendicular to the spread is represented by

$$\frac{x}{l} = \frac{y}{m} = \frac{z}{n} = \ldots\ldots = s.$$

Hence the intersection of this with (11) is at a distance $s$ from the origin, given by writing $ls$ for $x$, $ms$ for $y$ and so on, in (11), that is

$$l^2 s + m^2 s + n^2 s + \ldots\ldots = \frac{K}{\sqrt{A^2 + B^2 + C^2 + \ldots\ldots}}$$

But $$l^2 + m^2 + n^2 + \ldots\ldots = 1.$$

Hence the equation to a spread the perpendicular from the origin on which is in direction $(lmn\ldots)$ and of length $s$ is

$$lx + my + nz + \ldots\ldots = s \qquad (12).$$

Similarly we may show that if we have $q$ equations of the first degree representing a regular spread of the order $(p - q)$, that the directions whose coefficients are respectively proportional to the coefficients of the variables in these equations, are all perpendicular to the spread. And since we have $q$ equations which are independent and consistent, we shall get $q$ directions perpendicular to the spread, which may be shown to be independent directions, since the equations are independent equations. Thus we may have a spread of the $q^{\text{th}}$ order perpendicular to one of the order $(p - q)$.

Since the formulae we found in Chapter V. for the transformation of coordinates on moving the origin or rotating the axes in any way, are all formulae of the first degree in the variables, it follows that no movement of the axes can alter the degree of an equation representing a given spread. Hence the degree of an equation is an intrinsic property of the spread it represents. The obvious way therefore of investigating the properties of spreads represented by equations of a given degree is to investigate whether the equation cannot be simplified by moving the origin or axes. Thus in equations of the second degree we have terms such as $x^2$, and such as $xy$ of the second degree. It will be found that terms of the second class can always be got rid of by rotating the axes, and that sometimes terms of the first can, but that both classes can not be got rid of at once. Further, by shifting the origin, it is always possible to get rid of terms of the first degree if corresponding ones of the second degree remain, and if this is not the case it is still possible to get rid of all but one of the terms of the first degree, and of the constant term. In this way spreads of the second degree may be divided into those which can be reduced to terms of the second degree and a constant term only, and those which can not. As in the former class, if any point $(abcd...)$ is in the spread, the point $(-a-b-c-d...)$ is also in it, this class may be called central spreads, and the others non-central.

As this book does not profess to be a treatise on geometry of more than three independent directions, I will merely conclude by enumerating the chief characteristics of forms of the second degree, that is spreads represented by equations of the second degree in a space of four independent directions.

The equation to a central form may be reduced to the form

$$\pm \frac{x^2}{a^2} \pm \frac{y^2}{b^2} \pm \frac{z^2}{c^2} \pm \frac{w^2}{d^2} = 1. \quad\text{............ (13).}$$

If all the upper signs are taken it may be called the elliptic form. If $a = b = c = d$ the equation may be written

$$x^2 + y^2 + z^2 + w^2 = a^2 \quad\text{.................. (14),}$$

which, if the axes are rectangular, clearly represents a form every point in which is at a distance $a$ from the origin, that is, a circular form.

If one or more of the lower signs is taken in equation (13) we get a hyperbolic form. There are three orders of these. The first and third orders, that is the forms

$$\frac{x^2}{a^2} + \frac{y^2}{b^2} + \frac{z^2}{c^2} - \frac{w^2}{d^2} = \pm 1 \ldots\ldots\ldots(15)$$

are conjugate to each other. That is, they are asymptotic, both to each other, and to the form

$$\frac{x^2}{a^2} + \frac{y^2}{b^2} + \frac{z^2}{c^2} - \frac{w^2}{d^2} = 0 \ldots\ldots\ldots\ldots(16),$$

which I have called the ellipto-conical form.

The second order of hyperbolic forms are self conjugate, that is each of the two forms

$$\frac{x^2}{a^2} + \frac{y^2}{b^2} - \frac{z^2}{c^2} - \frac{w^2}{d^2} = \pm 1 \ldots\ldots\ldots(17)$$

are hyperbolic forms of the second order, and are conjugate to each other, and both asymptotic to the form

$$\frac{x^2}{a^2} + \frac{y^2}{b^2} - \frac{z^2}{c^2} - \frac{w^2}{d^2} = 0 \ldots\ldots\ldots\ldots(18),$$

which I have called the hyperbolo-conical form.

It may be shown that there are a number of straight lines, or generating lines as they may be called, wholly in the hyperbolic form of the first order, which do not however pass through every point in the form. The form of the second order may be completely generated by straight lines, as also may the ellipto-conical form. The hyperbolo-conical form may be generated similarly by planes, but there are neither generating planes nor lines in the elliptic form, or in the hyperbolic form of the third order.

The equation to a non-central form may be reduced to

$$\pm\frac{x^2}{a^2} \pm \frac{y^2}{b^2} \pm \frac{z^2}{c^2} + \frac{w}{d} = 0 \ldots\ldots\ldots (19).$$

If all the upper, or all the lower, signs are taken we have the ellipto-parabolic form; if the signs of the first three terms are not all the same, the hyperbolo-parabolic form. As the sign of the last term is indeterminate, these are the only material variations possible.

The hyperbolo-parabolic forms

$$\frac{x^2}{a^2} + \frac{y^2}{b^2} - \frac{z^2}{c^2} \pm \frac{w}{d} = 0 \dots\dots\dots\dots (20)$$

are conjugate, and asymptotic, both to each other, and to the cylindro-conical form

$$\frac{x^2}{a^2} + \frac{y^2}{b^2} - \frac{z^2}{c^2} = 0 \dots\dots\dots\dots\dots(20).$$

There are no generating lines in the ellipto-parabolic form, but the hyperbolo-parabolic may be completely generated by straight lines, and the cylindro-conical by planes. Besides these forms the equation may represent other cylindrical forms, when one or more coordinate vanishes from it, or it may reduce to one or two regular forms, to a single point, or to an impossible locus.

# CHAPTER VII.

We are now in a position to return to the main question—the objective truth or otherwise of my second axiom.

We have already granted that material space, the space in which material bodies are free to move, does extend from every position in it in a complete group of three independent directions. Consequently if it does not extend from all positions in it in the same directions, there must be some number, $p$, of independent directions, greater than three, on which all the directions in which it extends are dependent. Hence material space is a form of some kind, in a space of $p$ independent directions, and its shape is therefore represented by $(p-3)$ equations, in $p$ coordinates.

We may assume that the origin of the system of coordinates is in material space; that the axes are rectangular; and that the equations are all expressed in rational algebraical form.

Now there is one thing more that we know about material space, namely that it is a self-congruent space. From this it follows that from every position in material space its shape and size must appear the same, and therefore that if we move the origin to any point in material space it must always be possible, by a suitable rotation of the axes, to reduce the equation to it to the same form it had originally.

Now if we move the origin to any point $(x'y'z'...)$, to transform the equation we must write $(x + x')$ for $x$, $(y + y')$ for $y$, and so on, in all the equations. Hence if $x^n$; $x^m y^{n-m}$ be terms of the highest degree, $n$, in the old equations, these will become in the new equations,

$$(x^n + nx^{n-1}x' + ...)$$

and $\quad (x^n y^{n-m} + mx^{m-1}x'y^{n-m} + (n-m)x^m y'y^{n-m-1} + ...)$

and the only terms of the highest degree, $n$, will still be $x^n$, $x^m y^{n-m}$, as before. But terms of any lower degree will in general be altered. For each term of the degree $n$, produces terms of all degrees below $n$, and these cannot be cancelled by others arising from other terms for every possible movement of the origin. These terms will therefore have to be got rid of by rotating the axes, keeping them still rectangular, which may be done by the formulae (11) or (13) of Chapter V.

For the sake of simplicity we will suppose the formulae (11) for rotation of two axes at a time to be made use of repeatedly; that is, we write $(lx + my)$ for $x$, and $(mx - ly)$ for $y$ (where $l^2 + m^2 = 1$).

But, as the highest terms, of the degree $n$, were not changed by the translation of the origin, neither must they be changed by the rotation of the axes. Hence these terms must be of the form $A (x^2 + y^2)$ or some power of such an expression; that is $A (x^2 + y^2)^{\frac{n}{2}}$, so that $n$ must be an even number. But a similar rotation in the axis plane $(yz)$ shows that the terms of the highest order must also be included in the form

$$A (x^2 + y^2 + z^2)^{\frac{n}{2}},$$

and so on for all the coordinates. Hence each of the $(p - 3)$ equations representing material space must be of the form

$$A (x^2 + y^2 + z^2 \ldots)^{\frac{n}{2}} + \text{lower terms} = 0 \ldots\ldots\ldots\ldots (1),$$

where $A$ may in any one equation be zero, but if it is not, all those coordinates appear within the bracket which are capable of variation in material space. If more than one of the equations contain terms of the $n^{th}$ order, we may therefore eliminate such terms from all but one of them, by dividing each of them by their constant factor corresponding to $A$, and then subtracting them two and two. We have then one equation of the above form, and $(p - 4)$ of lower degrees.

If we now move the origin to another point in material space $(x'y'z'\ldots\ldots)$ the terms of the $n^{th}$ degree produce terms in all the variables of every degree below the $n^{th}$, and any other terms in the old equation also produce new terms of lower degrees. Since the old origin was in material space, the old equations contained no constant term, and the condition that the new origin is also in material space eliminates the constant term from the new equation. But there remain terms of all degrees up to the $(n-1)^{\text{th}}$, to be eliminated by the rotations of the axes. Now these rotations of two axes at a time afford us $\dfrac{q\,(q-1)}{1 \cdot 2}$ equations, if $q$ is the number of coordinates which are capable of variation in material space. Hence the rotation of the axes may be made to satisfy $\dfrac{q\,(q-1)}{1 \cdot 2}$ conditions, but not more. Now if $n = 2$ there are only terms of the first degree to be eliminated from the equations. And since only $q$ coordinates vary in material space, it must be possible to find constant values from the equations for the other $(p-q)$ which values must be zero since the origin is in material space. Hence there can not be more than $q$ terms of the first degree, and the $\dfrac{q\,(q-1)}{1 \cdot 2}$ conditions will be sufficient to bring them back to what they were in the original equation. But if $n$ is greater than 2, even if it is 4, besides the $q$ terms of the first degree, there will be $\dfrac{q\,(q-1)}{1 \cdot 2}$ of the second, and yet more of the third degree. Hence it is impossible to bring them all back to what they were originally. Therefore $n$ cannot be greater than 2, and equation (1) may be written

$$x^2 + y^2 + z^2 + w^2 + \ldots\ldots + \text{terms of the first degree} \ldots\ldots$$
$$= 0 \ldots\ldots\ldots\ldots(2).$$

The remainder of the equations being of a degree below $n$ are of the first degree. As we have seen $(p-q)$ of them may be reduced to the form

$$u = 0, \ldots\ldots\ldots\ldots\ldots\ldots\ldots(3).$$

And now let us revolve the axes so that three of them, $x$, $y$, and $z$, are tangent to material space at the origin. Hence $\dfrac{dF}{dx}\,\dfrac{dF}{dy}$ and $\dfrac{dF}{dz}$ must vanish when all the coordinates are equated to zero. Therefore there can be no terms of the first degree in $x$, $y$, $z$ in the equations. Thus the equations are reduced to the form

$$\begin{aligned} x^2 + y^2 + z^2 + w^2 + v^2 + \ldots\ldots + aw + bv + \ldots\ldots = 0 &\quad \text{(1 equation)} \\ a_1 w + b_1 v + \ldots\ldots = 0 \quad\quad (q-4) \text{ equations} & \\ u = 0 \quad\quad (p-q) \text{ equations} & \end{aligned} \Bigg\} \ldots\ldots(4).$$

Now write $(lw + mv)$ for $w$ and $(mw - lv)$ for $v$, rotating the axes of $w$, $v$, in their own plane, and determine $\dfrac{l}{m}$ by equating the coefficient of $w$ in the second of the above equations to zero, that is by writing $\dfrac{l}{m} = -\dfrac{b_1}{a_1}$. Similarly rotate the next pair of axes so as to eliminate $w$ from the second equation. And as there are $(q-4)$ coordinates besides $w$ in these $(q-4)$ equations we may repeat this process $(q-4)$ times, and so eliminate $w$ from all the $(q-4)$ equations, and leave only $(q-4)$ variables in them. They then suffice to determine constant values for all these variables, which are obviously all zero, and so we have finally left the equations to material space

$$x^2 + y^2 + z^2 + w^2 + a'w = 0 \qquad \text{(one equation)}$$
$$v = 0,\ u = 0 \ldots\ldots (p-4) \text{ equations} \Big\}\ \ldots\ldots(5).$$

These $(p-4)$ equations represent a regular spread of four independent directions, through the origin. Hence we have nothing more to do with the $(p-4)$ independent directions in which these coordinates are measured, but only with the spread of four independent directions, and the first of the above equations.

Hence, let us move the origin to the point $\left(0,\ 0,\ 0, -\dfrac{a'}{2}\right)$, in the spread of four independent directions, this equation becomes

$$x^2 + y^2 + z^2 + w^2 = \frac{a'^2}{4} \ldots\ldots\ldots\ldots\ldots(6),$$

and consequently:—If Material space is not a regular form, the only alternative consistent with the objective facts we have granted is that it should be a Circular Form, and this in however many independent directions points in it might be conceived to vary.

Mr Chrystal, starting on the definition of a straight line merely as a self-congruent line, came to rather a different conclusion. He deduced a theory of hyperbolic space, which I have now shown to be inconsistent with the assumption that space is self-congruent, which he nevertheless assumes in his demonstration. This might indeed have been indirectly inferred from the proof given in Part I. of Euclid's 12th Axiom. But I have now further shown that there cannot be more than one kind of 'elliptic space,' of which Mr Chrystal discussed several, and that kind is that which he calls 'double elliptic space,' but which is not elliptic, but circular.

But there are observed facts which prove that, if material space is not a regular form, it is a circular one of such enormous radius that the minute fraction of it of which we know anything at all may without sensible error be considered a regular form, for all practical purposes.

We have already seen that if material space is a circular form of radius $\rho$, in any triangle

$$A + B + C - \pi = \frac{\Delta}{\rho^2}.$$

That is, the excess of a triangle varies as its area. Now the largest triangle whose angles we could measure would be one inscribed in the earth's orbit. And in such a triangle there cannot be any great excess, or astronomers would ere now have observed it. Therefore its area must be small compared with that of a great sphere. Much more so therefore must the area of those triangles be by which the distance of the sun from the earth is determined. If therefore, as an approximation, we treat these as plane triangles, we may assume that the distance of the sun has been correctly calculated, at some 91 millions of miles, say. To find the area of a triangle inscribed in the earth's orbit, we may, as an approximation, consider it a plane triangle, and the orbit as circular and of radius $r$. The area of an equilateral triangle will then be $\frac{3\sqrt{3}}{4} r^2$.

Hence the excess of the triangle is $\frac{3\sqrt{3}}{4} \left(\frac{r}{\rho}\right)^2$.

Hence if $\epsilon$ be the circular measure of the excess

$$\frac{\rho}{r} = \left\{ \frac{3\sqrt{3}}{4} \cdot \frac{1}{\epsilon} \right\}^{\frac{1}{2}},$$

and so even if so considerable an excess as 10″ could have been overlooked, the radius of material space must be at least 160 times that of the earth's orbit. As the largest mundane triangle whose angles have been measured in a trigonometrical survey has probably an area considerably less than that of an equilateral triangle inscribed in a circle of 50 miles' radius, the excess of such a triangle would only be $\left(\frac{50}{91,000,000}\right)^2$ of the excess of a triangle inscribed in the earth's orbit, a quite imperceptible quantity.

But there is another consideration which may assure us that the radius of material space must in all probability be very many times greater than this, if it is not infinite, that is if material space is not a regular form.

For a long while the fixed stars were believed to have no parallax, even with such a base as the diameter of the earth's orbit. On the assumption that material space is a circular form this might indeed be explained as being due to the excess of the triangles used to measure their distances; which would mean that they must all lie upon or near the equatorial sphere of which the solar system occupies one pole, and therefore be separated from us by a distance of about a quadrant of a great circle. Inexplicable as such a disposition of the stars in space might seem, it is rendered still more so when we consider that some stars have in recent years been found to have as much parallax as 2″, or even more. These stars must therefore be separated by a distance less than a quadrant from us. But in that case how is it that no stars have been found whose distances are a little more than a quadrant? For such stars would have equal but *negative* parallaxes, a phenomenon which has never been observed.

But if we reject this explanation of the smallness of the parallax of the fixed stars, assuming, as seems so much more probable, that they are more or less evenly distributed throughout space, then the most distant of them which is visible to us must be many times as distant from us as the nearest, and its distance is not greater than a quadrant of a great circle. Hence the triangles used to determine the parallax of the nearest fixed stars must be approximately plane, and their calculated distances therefore approximately correct. Now the distance of a star with 2″ parallax would be about 100,000 times the diameter of the earth's orbit, and consequently a quadrant of circular space must be many times this—that is, millions of times the distance from us to the sun. Thus we cannot hope to discover any excess, even in a triangle inscribed in the earth's orbit!

Now no objective knowledge can attain more decisive results than this. The inductions on the strength of which we accepted my other two axioms as objectively true of material space, are certainly no stronger, and before doubting the objective truth of the remaining one it would therefore be necessary to reconsider them. We have then come to the following

## CONCLUSIONS.

I. THERE EXISTS A SUBJECTIVE GEOMETRY, WHOSE SUBJECTIVE CONCLUSIONS ARE NECESSARY TRUTHS.

II. THAT THE CONCLUSIONS OF THIS GEOMETRY ARE ALSO APPLICABLE TO THE OBJECTIVE GEOMETRY OF MATERIAL SPACE, IS PROVED BY INDUCTIONS AS CONVINCING AS ANY WE KNOW OF, EXCEPT PERHAPS THAT WHICH CONVINCES US THAT THERE IS AN OBJECTIVE UNIVERSE AT ALL.

QUOD ERAT DEMONSTRANDUM.

THE END.

CAMBRIDGE: PRINTED BY C. J. CLAY, M.A. AND SONS, AT THE UNIVERSITY PRESS.

www.ingramcontent.com/pod-product-compliance
Lightning Source LLC
Chambersburg PA
CBHW021815190326
41518CB00007B/599